Multiple Models Approach in Automation

Multiple Models Approach in Automation

Takagi-Sugeno Fuzzy Systems

Mohammed Chadli
Pierre Borne

Series Editor
Bernard Dubuisson

iSTE

WILEY

First published 2013 in Great Britain and the United States by ISTE Ltd and John Wiley & Sons, Inc.

ISTE Ltd
27-37 St George's Road
London SW19 4EU
UK

www.iste.co.uk

John Wiley & Sons, Inc.
111 River Street
Hoboken, NJ 07030
USA

www.wiley.com

Library of Congress Control Number: 2012950089

British Library Cataloguing-in-Publication Data
A CIP record for this book is available from the British Library
ISBN: 978-1-84821-412-5

Table of Contents

Notations

Notations relating to multiple modeling

PDC	*Parallel distributed compensation* (control law based on state feedback)
OPDC	*Output PDC* (control law based on output feedback)
$(x(\cdot), u(\cdot), y(\cdot))$	State, input and output of the system, respectively, $(x(\cdot), u(\cdot), y(\cdot)) \in \mathbb{R}^n \times \mathbb{R}^m \times \mathbb{R}^l$
(A_i, B_i, C_i)	State, input and output matrices of the ith local LTI model, such that $(A_i, B_i, C_i) \in \mathbb{R}^{n.n} \times \mathbb{R}^{n.m} \times \mathbb{R}^{l.n}$
$\mu_i(\cdot)$	ith activation function corresponding to the ith local LTI model, such that $\mu_i(\cdot) \geq 0, \sum_{i=1}^{N} \mu_i(\cdot) = 1$
N	Number of local LTI models
$\sum_{i \neq j:1}^{N} x_i x_j$	This expression represents $\sum_{i=1}^{N} \sum_{j=1, i \neq j}^{N} x_i x_j$
$\sum_{i < j:1}^{N} x_i x_j$	This expression represents $\sum_{i=1}^{N} \sum_{j=1, i < j}^{N} x_i x_j$

Sets and domains

$\mathbb{R}$	Set of real numbers
$\mathbb{R}^+ (\mathbb{R}^{+*})$	Set of positive (or strictly positive, respectively) or null real numbers
$\mathbb{R}^p$	Euclidean real space of size p
$\mathbb{R}^{p.q}$	Set of real element matrices of size $(p.q)$
$\mathbb{C}$	Set of complex numbers
I_n	Set of natural whole numbers ranging from 1 to n $I_n = \{1, 2, .., n\}$
C^1	Set of continuously differentiable functions

Matrices and operators

$M > 0 (M \geq 0)$	Positive definite (or semi-definite, respectively) symmetric square matrix
$M < 0 (M \leq 0)$	Negative definite (or semi-definite, respectively) symmetric square matrix
$M < N$	Negative definite symmetric $M - N < 0$ square matrix
$I_p (I)$	Identity matrix of size p (or appropriate size, respectively)
$\det(M)$	Determinant of square matrix M
$\text{rang}(M)$	Rank of matrix M
M^{-1}	Inverse of square matrix M
$M^T (M^{-T})$	Transpose of M and of the inverse of M, respectively
M^-	Moore–Penrose generalized inverse of M
$M_\perp$	Matrix Z such that $MZ = 0$ and matrix $(M^T \; M_\perp)$ of maximum rank
$\begin{pmatrix} M_{11} & M_{12} \\ * & M_{22} \end{pmatrix}$	Symmetric matrix, symbol $*$ represents M_{12}^T
$\lambda(M)$	Eigenvalues of matrix M
$\lambda_{\min}(M)$	Minimum eigenvalue of matrix M
$\lambda_{\max}(M)$	Maximum eigenvalue of matrix M
$\Re(\lambda(M))$	Real parts of eigenvalues of matrix M
$\Im(\lambda(M))$	Imaginary parts of eigenvalues of matrix M
$\|M\|$	Euclidean norm of square matrix M

$\|x\|$	Euclidean norm of vector x of size p, $\|x\|^2 = \sum_{i:1}^{p} x_i^2$
$\langle x, y \rangle$	Scalar product, $\langle x, y \rangle = x^T y, (x, y) \in \mathbb{R}^p \times \mathbb{R}^p$
$\otimes$	Kronecker product

Acronyms

LMI	Linear matrix inequality
BMI	Bilinear matrix inequality
LTI	Linear time invariant
GEVP	Generalized eigenvalues problem
PLDI	Polytopic linear differential inclusion
MIMO	Multiple input, multiple output
SIMO	Single input, multiple output

Introduction

In recent decades, many studies on analysis and synthesis problems relating to the multiple model (also called multimodels) approach have been undertaken. This has been motivated by the desire to establish the design problems in numerical terms. These have become possible as a result of the convex polytopic representation of the multiple model approach and the development of effective numerical resolution tools based on convex optimization software.

Automation relies on the concept of a model representing the internal dynamic behavior of a system. For example, a system is modeled through a mathematical relationship of input/output behaviors, or through an equation relating to its change of state. The dilemma is then to choose between the reliability of the model using the real process and the adequacy of this model expressed in mathematical form. Faced with this problem, automation engineers are often led, based on physical considerations, to consider certain classes of systems according to structural restrictions (such as linearity and convexity), leading to model approximations. Thus, linear representation (LTI models) has been widely used. The nonlinear model is thus represented by a single linear model, obtained from first approximation and close to an operating point. The disadvantage of such an approach is its local aspect, the linear model only being a local description of system behavior.

In recent years, a general approach based on multiple LTI models (linear or affine) around various function points has been proposed. This so-called multiple model (also called multimodel or Takangi-Sugeno fuzzy model) approach is a convex polytopic representation, which can be obtained either directly from a nonlinear mathematical model, through mathematical transformation or through linearization around various operating points. Many studies relating to the stability of this class of nonlinear systems have been published in recent years. Initially, these works were inspired by linear system control techniques, leading to the use of studies following quadratic and nonquadratic Lyapunov approach. Numerous approaches have been developed to study the stability of different system categories, such as uncertain, nonlinear, bilinear, varying parameter and delayed systems.

This book deals with the stability analysis and synthesis of control laws and observers for multiple models. Our approach is essentially based on Lyapunov's second method and LMI formulation. Uncertain multiple models with unknown inputs are also studied. Quadratic and nonquadratic Lyapunov functions are considered. Interest in the quadratic method stems from the fact that it is easy to use the synthesis search parameters, which can be set out as a convex optimization problem in LMI form. However, the quadratic method has turned out to be very conservative, in that this approach ignores all of the data contained in the activation functions. These constraints become still more conservative if we add performance constraints of the closed-loop system.

In order to increase confidence in the quadratic method, multiple model stability studies are carried out by considering nonquadratic Lyapunov functions.

This book consists of five chapters dealing exclusively with continuous-time multimodels. It is set out as follows:

– Chapter 1 provides an introduction to multiple model representation and the tools used. Various methods used to obtain a multiple model, LMI tools and various control laws are presented.

– Chapter 2 presents different stability conditions. Using quadratic and nonquadratic Lyapunov functions, suitable conditions for stability are proposed for multiple models with uncertain parameters.

– Chapter 3 is integrally dedicated to observer synthesis. Unmeasurable decision variables and the separation principle are tackled. State estimation in the presence of unknown and uncertain inputs is covered. Various techniques are examined and LMI synthesis conditions are proposed. Some illustrative examples are included.

– Chapters 4 and 5 deal with stabilization through full state and output feedback controllers. The conditions obtained are expressed in the bilinear form (BMI), then LMI design conditions are proposed. Closed-loop multiple model performance is ensured through the placement of poles in LMI regions. The synthesis of robust control laws is also tackled by considering two types of parametric uncertainty (structured and interval uncertainties).

Various examples are proposed in order to illustrate the theoretical developments.

Chapter 1

Multiple Model Representation

1.1. Introduction

Following the work of Zadeh [ZAD 65], there has been a high degree of success in the use of fuzzy logic in the modeling of complex/nonlinear systems and also in the synthesis of fuzzy controllers [TAK 85]. The ability of fuzzy logic to represent a wide class of systems has been demonstrated as a universal approximation. In this respect, a number of successful applications have been achieved [BUC 93, CAS 95]. Various fuzzy models can be found in literature. However, two principal models have come to light: the Mamdani and Takagi–Sugeno (T-S) [TAK 85] models. Mamdani's fuzzy model uses fuzzy subsets for the most part, whereas the T-S type uses functions which are dependent on input variables. The most popular T-S model is the one which mostly uses a state–space or autoregressive model. This type of representation, known as multiple model representation [MUR 97], has been successfully used in all areas of automation (such as identification, control, FDI and FTC) [AKH 07a, AKH 07b, CHA 10a, CHA 09, CHA 08b, FRA 90, PAT 97].

1.2. Techniques for obtaining multiple models

Multiple models are obtained by interpolation between linear time invariant (LTI) models. Each LTI model represents an operating range which is valid around an operating point. Three methods are used to obtain a multiple model:

– by identification when input and output data is available;

– by linearization around various operating points;

– by a convex polytopic transformation when an analytical model is available.

These models use state–space representation. Thus, studies dealing with the stability analysis of multiple models as well as synthesis of controllers and observers adopt state–space representation in order to extend to nonlinear systems, results widely used in the linear domain.

Continuous-time and discrete-time multiple models, are generally of the form:

$$\dot{x}(t) = \sum_{i=1}^{N} \mu_i(z(t))(A_i x(t) + B_i u(t)) \qquad [1.1]$$

$$x(t+1) = \sum_{i=1}^{N} \mu_i(z(t))(A_i x(t) + B_i u(t)) \qquad [1.2]$$

where $x(\cdot) \in \mathbb{R}^n$ is the state variables vector, $u(\cdot) \in \mathbb{R}^m$ the input vector and $z(\cdot) \in \mathbb{R}^q$ the decision variables vector. $\mu_i(\cdot)$, $i \in I_N$, represent activation functions such that $\sum_{i=1}^{N} \mu_i(z(t)) = 1$, $\mu_i(z(t)) \geq 0$.

1.2.1. *Construction of multiple models by identification*

"Black box" models are identified from inputs and outputs data around various operating points. Independently of the type of chosen model, this identification requires an "optimal" structure to be found, as well as an estimation of parameters and a validation of the final model [GAS 01, JOH 00, JOH 03, MUR 97, TAK 85]. In our case, the model is nonlinear relative to the parameters. Some iterative nonlinear optimization techniques are used according to the available data *a priori*. Identification methods for the unknown parameters are generally based on the minimization of a functional of the difference between the estimated output of the multiple model $y_m(t)$ and the measured output of the system $y(t)$. The criterion commonly used is the minimization of the quadratic error:

$$J(\theta)=\frac{1}{2}\sum_{k=1}^{H}\varepsilon^2(k,\theta)=\frac{1}{2}\sum_{k=1}^{H}\left(y_m(k)-y(k)\right)^2 \qquad [1.3]$$

where H is the observation time and θ is the parameters vector (LTI models and activation functions).

The algorithm is updated using the following general equation:

$$\theta(k+1)=\theta(k)-\eta D(k) \qquad [1.4]$$

where k is the iteration index, $\theta(k)$ is the value for the solution at iteration k and η represents an adjustment factor calculated using a heuristic method, which allows us to regulate the rate of convergence toward the solution. $D(k)$ is the research direction, for which the expression is dependent on the optimization method used. The following are the three most frequently used optimization methods:

– Gradient algorithm, which is based on a first-order development of criterion $J(\theta)$:

$$D(k) = G(\theta(k)) = \left.\frac{\partial J}{\partial \theta}\right|_{\theta=\theta(k)} = \left.\sum_{k=1}^{H} \frac{\partial \varepsilon(k,\theta)}{\partial \theta} \varepsilon(k,\theta)\right|_{\theta=\theta(k)} \quad [1.5]$$

– Newton algorithm, which is based on a second-order development:

$$D(k) = H^{-1}(k) G(\theta(k)) \quad [1.6]$$

where $H(k)$ is the Hessian matrix, defined by:

$$H(k) = \left.\sum_{k=1}^{H} \frac{\partial \varepsilon(k,\theta)}{\partial \theta} \frac{\partial \varepsilon(k,\theta)}{\partial \theta^T} + \sum_{k=1}^{H} \frac{\partial^2 \varepsilon(k,\theta)}{\partial \theta^2} \varepsilon(k,\theta)\right|_{\theta=\theta(k)} \quad [1.7]$$

The advantage of this algorithm is in its ability to simultaneously define the direction and speed of research. Its main disadvantage is the inverse Hessian calculation at each iteration.

– Gauss–Newton algorithm, which is a simplified form of the Newton algorithm uses an approximation of the Hessian ignoring the second-order terms:

$$H_a = \sum_{k=1}^{H} \frac{\partial \varepsilon(k,\theta)}{\partial \theta} \frac{\partial \varepsilon(k,\theta)}{\partial \theta^T} \quad [1.8]$$

This algorithm guarantees a positive definite Hessian and consequently convergence toward a minimum.

These algorithms are sensitive to the initial choice of parameter vector θ and when the size of the parameter space is highly significant, the risk is that the algorithms will converge toward local minima. More detailed information on this topic can be found in, for example [BOR 90, BOR 92].

EXAMPLE 1.1.– In order to illustrate the method, let us consider the following four-wheel vehicle model [CHA 07b]:

$$\begin{pmatrix} \dot{\beta}(t) \\ \dot{r}(t) \end{pmatrix} = \begin{pmatrix} 2\dfrac{F_f(t)+F_r(t)}{mU} - r(t) \\ 2\dfrac{a_f F_f(t) - a_r F_r(t)}{I_z} \end{pmatrix} \qquad [1.9]$$

where β is the lateral slip angle, r is the yaw rate angle, F_f and F_r are the front and rear pneumatic forces, respectively, U is the vehicle speed, assumed to be constant, I_z is the moment of inertia relative to the z axis and m is the mass of the vehicle.

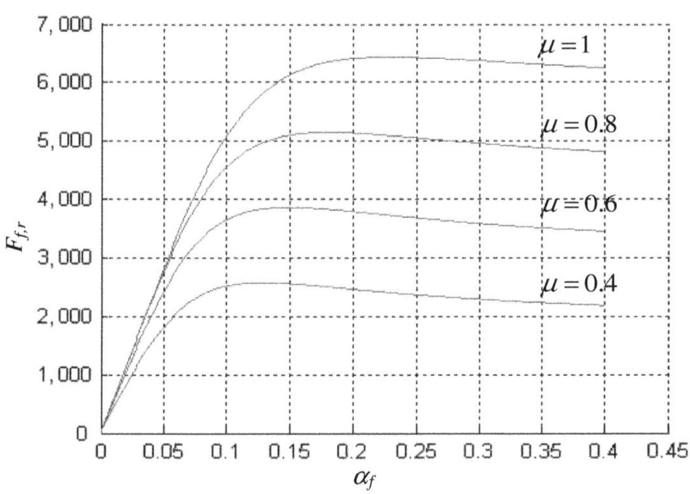

Figure 1.1. *Nonlinear behavior of contact forces*

Various pneumatic force models exist in literature. Among these, the most used is that expressed in terms of tire slip angles by [CHA 07b]:

$$F_f = D_f \sin\left(C_f \tan^{-1}\left(\beta_f(1-E_f)\alpha_f + E_f \tan^{-1}(\beta_f\alpha_f)\right)\right) \quad [1.10]$$

$$F_r = D_r \sin\left(C_r \tan^{-1}\left(\beta_r(1-E_r)\alpha_r + E_r \tan^{-1}(\beta_r\alpha_r)\right)\right) \quad [1.11]$$

Our aim is to approximate these front and rear tire-to-road contact forces (***F_f*** and ***F_r***) ([1.10] and [1.11]) using a multiple model depending on tire slip angle $\alpha_i\,(i = f,r)$ and adhesion coefficient μ by:

$$\begin{cases} F_f = h_1(|\alpha_f|)C_{f1}(\mu)\alpha_f + h_2(|\alpha_f|)C_{f2}(\mu)\alpha_f \\ F_r = h_1(|\alpha_f|)C_{r1}(\mu)\alpha_r + h_2(|\alpha_f|)C_{r2}(\mu)\alpha_r \end{cases} \quad [1.12]$$

where C_{fi} and C_{ri} represent the stiffness coefficients which are dependent on adhesion coefficient μ. The activation functions h_i $(\mathrm{i} = 1,2)$ satisfy the following properties:

$$\sum_{i=1}^{2} h_i(|\alpha_f|) = 1, \quad 0 \le h_i(|\alpha_f|) \le 1, \; i = 1,2 \quad [1.13]$$

By choosing activation functions of the form:

$$h_1(|\alpha_f|) = \frac{\omega_1(|\alpha_f|)}{\omega_1(|\alpha_f|) + \omega_2(|\alpha_f|)} = 1 - h_2(|\alpha_f|) \quad [1.14]$$

where

$$\omega_1(|\alpha_f|) = \frac{1}{\left(1 + abs\left(\frac{|\alpha_f| - c_1}{a_1}\right)\right)^{2b_1}},$$

$$\omega_2(|\alpha_f|) = \frac{1}{\left(1 + abs\left(\frac{|\alpha_f| - c_2}{a_2}\right)\right)^{2b_2}} \quad [1.15]$$

and by using an algorithm such that as those defined above, for $\mu = 0.7$, the following values are obtained:

Nominal stiffness coefficients	C_{f1}	C_{f2}	C_{r1}	C_{r2}
Values	**60,712**	**4,812**	**60,088**	**3,455**

and the following activation function parameter values:

$$a_1 = 0.5077, b_1 = 0.4748, c_1 = 3.1893, a_2 = 5.3907, \\ b_2 = 0.4356, c_2 = 0.5633 \qquad [1.16]$$

Figure 1.2 allows us to compare the two models: the model described in [1.10] and [1.11] and the proposed multiple model. This model will be used in later chapters for the controller synthesis of the vehicle model [1.9].

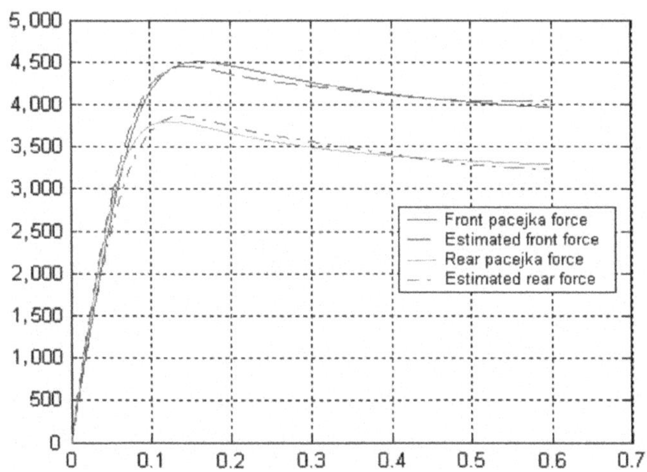

Figure 1.2. *Approximation of forces [1.10] and [1.11] for $\mu = 0.7$ using a multiple model*

1.2.2. *Multiple model construction by linearization*

In this example, the availability of a nonlinear mathematical model of a physical process that can be linearized around various judiciously chosen operating points is assumed. Let us consider the following nonlinear system:

$$\dot{x}(t) = f\left(x(t), u(t)\right) \qquad [1.17]$$

where $f(x) \in C^1, f : \mathbb{R}^n \longrightarrow \mathbb{R}^n$. The linearization of system [1.17] around an arbitrary operating point $(x_i, u_i) \in \mathbb{R}^n \times \mathbb{R}^m$ is:

$$\dot{x}(t) = A_i(x\ (t) - x_i) + B_i(u(t) - u_i) + f(x_i, u_i) \qquad [1.18]$$

which can be rewritten in the form:

$$\dot{x}(t) = A_i x(t) + B_i u(t) + d_i \qquad [1.19]$$

where:

$$A_i = \left.\frac{\partial f(x,u)}{\partial x}\right|_{\substack{x=x_i \\ u=u_i}}, \quad B_i = \left.\frac{\partial f(x,u)}{\partial u}\right|_{\substack{x=x_i \\ u=u_i}} \qquad [1.20]$$

$$d_i = f\left(x_i, u_i\right) - A_i x_i - B_i u_i \qquad [1.21]$$

By assuming that the local models result from a linearization around N operating points (x_i, u_i), multiple model formulation leads to:

$$\dot{x}(t) = \sum_{i=1}^{N} \mu_i\left(z(t)\right)\left(A_i x(t) + B_i u(t) + d_i\right) \qquad [1.22]$$

where $\mu_i(z(t))$, $i \in I_N$, refers to the activation functions and $z(t)$ is the decision variables vector, which is dependent on measurable state variables and eventually on control $u(t)$. We note that in this case, the number of local models (N) is dependent on the desired precision of the modelization, the complexity of the nonlinear system and on the choice of activation functions structure.

EXAMPLE 1.2.– Let us consider the vehicle model which couples longitudinal and lateral dynamics [AKH 07a]:

$$
\begin{aligned}
u(t) &= v(t)r(t) - fg + \frac{(fk_1 - k_2)}{M}u(t)^2 + c_f\frac{v(t)+ar(t)}{Mu(t)}\delta(t) + \frac{T(t)}{M} \\
v(t) &= -u(t)r(t) - \frac{(c_f - c_r)}{Mu(t)}v(t)^2 + \frac{(bc_r - ac_f)}{Mu(t)}r(t) + \frac{(c_f\delta(t) + T\delta(t))}{M} \\
r(t) &= \frac{(bc_r - dc_f)}{I_z u(t)}v(t) - \frac{(b^2c_r - a^2c_f)}{I_z u(t)}r(t) + \frac{(aT(t)\delta(t) + ac_f\delta(t))}{I_z}
\end{aligned}
\quad [1.23]
$$

where $u(t)$, $v(t)$ and $r(t)$ are the longitudinal and lateral velocities and yaw rate respectively, $\delta(t)$ is the steering angle and $T(t)$ is the traction/turning force. Other vehicle parameters are given in Table 1.1.

M	Vehicle mass (1,480 kg)
I_z	Moment of inertia (2,350 kg m^2)
g	Gravitational force (9.81 m/s^2)
f	Rotational friction coefficient (0.02)
a	Distance from front axle to center of gravity (1.05 m)
b	Distance from rear axle to center of gravity (1.63 m)
c_f	Adhesion coefficient from front tires (1,35,000 N/rad)
c_r	Adhesion coefficient from rear tires (95,000 N/rad)

Table 1.1. *Model parameters*

The vehicle model is nonlinear and of the form $\dot{x}(t) = f(x(t), w(t))$, where $x(t) = [u(t), v(t), r(t)]^T$, $w(t) = [\delta(t), T(t)]^T$. Our aim is to approximate this model using the T-S model of the form [1.22], i.e.:

$$\dot{x}(t) = \sum_{i=1}^{N} \mu_i(z(t))\left(A_i x(t) + B_i u(t) + d_i\right) \qquad [1.24]$$

where:

$$\sum_{i=1}^{N} \mu_i(z(t)) = 1,\ \ \mu_i(z(t)) \geq 0$$

and

$$A_i = \left.\frac{\partial f(x,w)}{\partial x}\right|_{\substack{x=x_i \\ w=w_i}},\ B_i = \left.\frac{\partial f(x,w)}{\partial u}\right|_{\substack{x=x_i \\ w=w_i}},\ d_i = f(x_i, w_i) - A_i x_i - B_i w_i \qquad [1.25]$$

$$A_i = \begin{bmatrix} \dfrac{2(fk_1 - k_2)u_i}{M} - \dfrac{c_f(v_i + ar_i)}{Mu_i^2}\delta_i & r_i + \dfrac{c_f \delta_i}{Mu_i} & v_i + \dfrac{ac_f \delta_i}{Mu_i} \\ -r_i + \dfrac{(c_f + c_r)}{Mu_i^2} v_i - \dfrac{(bc_f - ac_f)}{Mu_i^2} r_i & \dfrac{(c_f + c_r)}{Mu_i^2} & u_i + \dfrac{(bc_r - ac_f)}{Mu_i^2} \\ -\dfrac{(bc_r - ac_f)}{I_z u_i^2} v_i + \dfrac{(b^2 c_r - a^2 c_f)}{I_z u_i^2} r_i & \dfrac{(bc_r - ac_f)}{I_z u_i^2} & -\dfrac{(b^2 c_r - a^2 c_f)}{I_z u_i^2} \end{bmatrix} \qquad [1.26]$$

$$B_i = \begin{bmatrix} c_f \dfrac{v_i + ar_i}{Mu_i} & \dfrac{1}{M} \\ \dfrac{c_f + T_i}{M} & \dfrac{\delta_i}{M} \\ \dfrac{aT + ac_f}{I_z} & \dfrac{a\delta_i}{I_z} \end{bmatrix} \qquad [1.27]$$

$$D_i = \begin{bmatrix} v_i r_i - fg + \dfrac{fk_1 - k_2}{M} u_i^2 + c_f \dfrac{v_i + ar_i}{Mu_i} \delta_i + \dfrac{T_i}{M} \\ -u_i r - \dfrac{c_f + c_r}{Mu_i} v_i + \dfrac{bc_r - ac_f}{Mu_i} r_i + \dfrac{c_f + T_i}{M} \delta_i \\ \dfrac{bc_r - ac_f}{I_z u_i} v_i - \dfrac{b^2 c_r - a^2 c_f}{I_z u_i^2} r_i + \dfrac{aT + ac_f}{I_z} \delta_i \end{bmatrix} - A_i \begin{bmatrix} u_i \\ v_i \\ r_i \end{bmatrix} - B_i \begin{bmatrix} \delta_i \\ T_i \end{bmatrix}$$

[1.28]

By using an algorithm such as that above, three local models have been chosen, which allows us to reach a good compromise between the precision and complexity of the model. The chosen activation functions are triangular and are dependent on longitudinal velocity $u(t)$ (Figure 1.5). The numerical values obtained from the different matrices A_i, B_i, D_i are:

$$A_1 = \begin{bmatrix} 0.052 & 0.403 & 0.239 \\ -0.366 & -10.82 & -13.743 \\ 0.728 & 0.388 & -11.890 \end{bmatrix}, B_1 = \begin{bmatrix} 10.99 & 7.10^{-4} \\ 91.216 & 10^{-4} \\ 60.319 & 0 \end{bmatrix}, D_1 = \begin{bmatrix} -0.832 \\ 5.259 \\ -10.46 \end{bmatrix}$$

$$A_2 = \begin{bmatrix} 0.085 & 2.895 & 1.925 \\ -0.989 & -9.282 & -16.213 \\ 0.507 & 0.333 & -10.198 \end{bmatrix}, B_2 = \begin{bmatrix} 3.359 & 7.10^{-4} \\ 91.216 & 3.10^{-4} \\ 60.319 & 2.10^{-4} \end{bmatrix}, D_1 = \begin{bmatrix} -0.087 \\ 16.562 \\ -8.496 \end{bmatrix}$$

$$A_3 = \begin{bmatrix} 0.031 & 2.065 & 0.693 \\ -1.141 & -8.468 & -17.870 \\ 0.441 & 0.303 & -9.303 \end{bmatrix}, B_3 = \begin{bmatrix} 1.548 & 7.10^{-4} \\ 91.216 & 2.10^{-4} \\ 60.319 & 10^{-4} \end{bmatrix}, D_1 = \begin{bmatrix} -0.392 \\ 20.951 \\ -8.092 \end{bmatrix}$$

[1.29]

Figures 1.6–1.8 show the approximation quality with the same control signals (Figures 1.3–1.4).

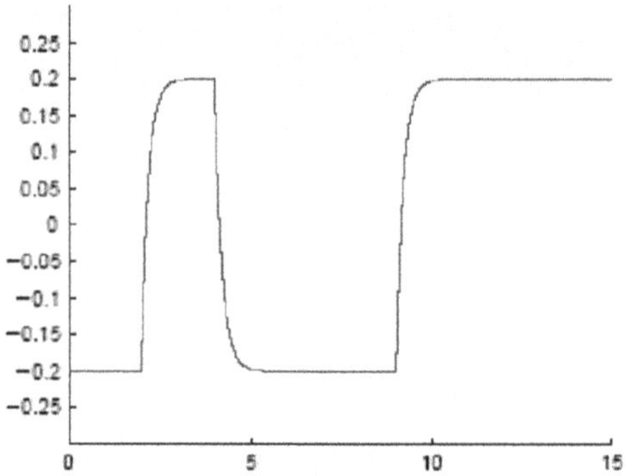

Figure 1.3. *Steering angle*

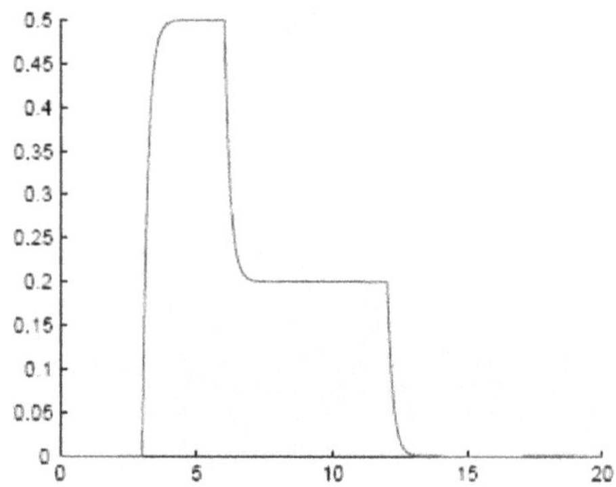

Figure 1.4. *Traction force*

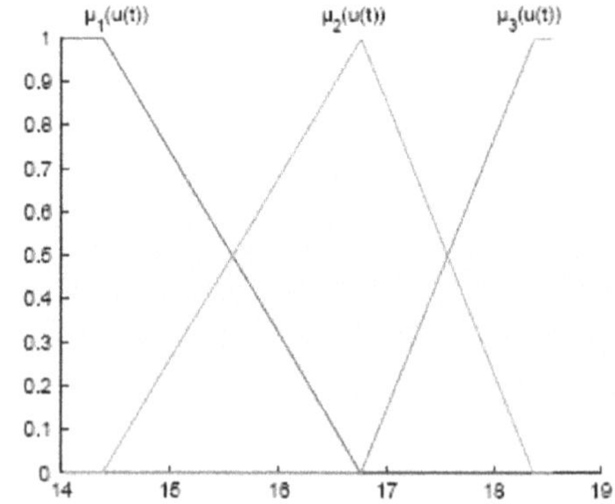

Figure 1.5. *Activation functions*

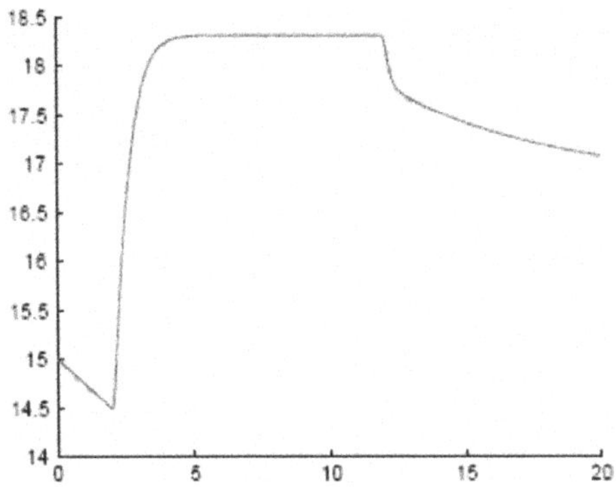

Figure 1.6. *Longitudinal velocity and its estimate*

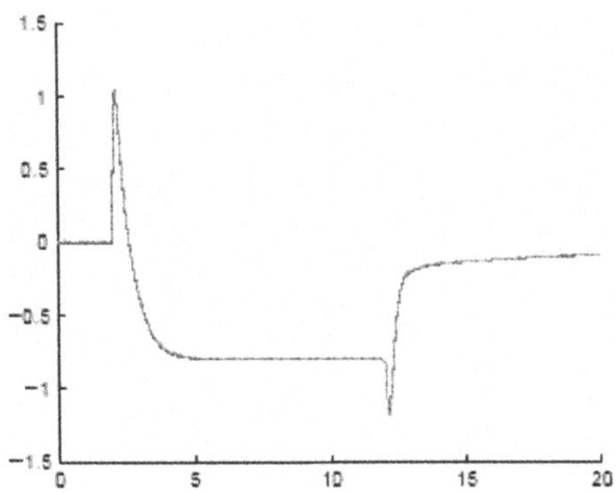

Figure 1.7. *Lateral velocity and its estimate*

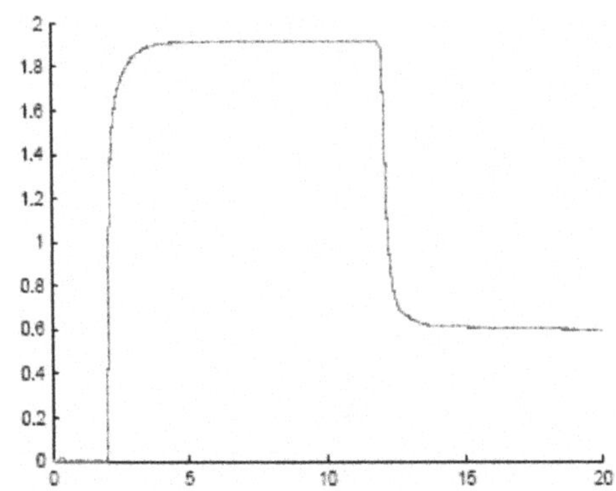

Figure 1.8. *Roll rate and its estimate*

1.2.3. *Multiple model construction by mathematical transformation*

Let us consider the following nonlinear system:

$$\begin{cases} \dot{x}(t) = f\left(x(t)\right) + B\left(x(t)\right)u(t) \\ y(t) = g\left(x(t)\right) + D(t)u(t) \end{cases} \qquad [1.30]$$

where $x(.) \in \mathbb{R}^n$, $u(.) \in \mathbb{R}^m$, $y(.) \in \mathbb{R}^l$, $f\left(x(.)\right) \in \mathbb{R}^n$, $g\left(x(.)\right) \in \mathbb{R}^l$, $B\left(x(.)\right) \in \mathbb{R}^{n.m}$ and $D\left(x(.)\right) \in \mathbb{R}^{l.m}$.

The proposed method uses only the boundedness of nonlinear terms [CHA 02b, MOR 00]. This method is based on a convex polytopic transformation of nonlinear scalar functions. It allows us to reduce the number of LTI models to two models. Indeed, for a bounded function $h\left(x(t)\right)$: $[a,b] \to \mathbb{R} \quad \forall \quad x \in [a,b]$, where $(a,b) \in \mathbb{R}^2$, there are two functions:

$$F^i(.):[a,b] \to [0,1], i \in I_2 \qquad [1.31]$$

$$x(t) \mapsto F^i\left(x(t)\right)$$

where $F^1\left(x(t)\right) + F^2\left(x(t)\right) = 1$ and two scalars α and β such that:

$$h\left(x(t)\right) = F^1\left(x(t)\right)\alpha + F^2\left(x(t)\right)\beta \qquad [1.32]$$

An evident decomposition of $h\left(x(t)\right)$ is to consider over $[a,b]$:

$$\beta \le h\left(x(t)\right) \le \alpha \qquad [1.33]$$

where:

$$\beta = \min_{x\in[a,b]}\left(h\left(x(t)\right)\right), \alpha = \max_{x\in[a,b]}\left(h\left(x(t)\right)\right) \quad [1.34]$$

$$F^1\left(x(t)\right) = \frac{h\left(x(t)\right)-\beta}{\alpha-\beta}, F^2\left(x(t)\right) = \frac{\alpha-h\left(x(t)\right)}{\alpha-\beta} \quad [1.35]$$

This decomposition method, which is not unique, will be used later.

Effectively, model [1.8] can be expressed by:

$$\begin{pmatrix} \dot{x}(t) \\ y(t) \end{pmatrix} = \begin{pmatrix} A\left(x(t)\right) & B\left(x(t)\right) \\ C\left(x(t)\right) & D\left(x(t)\right) \end{pmatrix} \begin{pmatrix} x(t) \\ u(t) \end{pmatrix} \quad [1.36]$$

and define:

$$E\left(x(t)\right) = \begin{pmatrix} A\left(x(t)\right) & B\left(x(t)\right) \\ C\left(x(t)\right) & D\left(x(t)\right) \end{pmatrix} \quad [1.37]$$

On the assumption that $E\left(x(t)\right)$ is continuous and bounded and considering each nonconstant constant, matrix $E\left(x(t)\right)$ can be transformed as follows:

$$E\left(x(t)\right) = \sum_{i=1}^{N} \mu_i\left(x(t)\right) E_i,\ E_i = \begin{pmatrix} A_i & B_i \\ C_i & D_i \end{pmatrix} \quad [1.38]$$

The number of local models N resulting from the transformation is consequently dependent on the number s of nonlinearities contained in matrix $E(x)$. This number is equal to $N = 2^s$. Some illustrative examples of this method are proposed later.

The advantage of this method is that no approximation errors are introduced and the number of local models is reduced compared to the linearization method. Remember that the reduction of the number of LTI models allows us to reduce the number of constraints (relating to stability or stabilization), which increases the chances of finding a solution.

EXAMPLE 1.3.– Let us consider the following nonlinear model:

$$\begin{aligned} \dot{x}(t) &= A(x(t))x(t) + Bu(t) \\ y(t) &= Cx(t) \end{aligned} \tag{1.39a}$$

where:

$$x(t) = \begin{pmatrix} x_1(t) \\ x_2(t) \end{pmatrix}, \; A(x) = \begin{pmatrix} -1 & \sin(x_2(t)) \\ 2 & -3 \end{pmatrix}$$
$$B = \begin{pmatrix} 1 \\ 1 \end{pmatrix}, \; C = \begin{pmatrix} 0 & 2 \end{pmatrix} \tag{1.39b}$$

The proposed transformation method is based on the boundedness of nonlinear continuous terms. Indeed, the nonlinear term $h_1(x_2(t)) = \sin(x_2(t))$ is bounded $\forall \; x(t) \in \mathbb{R}^n$:

$$-1 \le h_1(x_2(t)) \le 1$$

Thus, this nonlinear term can be transformed as follows:

$$h_1(x_2(t)) = -1.\mu_1(x_2) + 1.\mu_2(x_2)$$

where:

$$\mu_1(x_2) + \mu_2(x_2) = 1$$

which enables us to obtain:

$$\mu_1(x_2) = \frac{1-\sin(x_2)}{2}, \; \mu_2(x_2) = \frac{1+\sin(x_2)}{2}$$

Therefore:

$$\dot{x}(t) = \sum_{i=1}^{N=2} \mu_i(x(t))\, A_i x(t) + Bu(t)$$
$$y(t) = Cx(t)$$

where:

$$A_1 = \begin{pmatrix} -1 & -1 \\ 2 & -3 \end{pmatrix}, \; A_2 = \begin{pmatrix} -1 & 1 \\ 2 & -3 \end{pmatrix}$$

Now, if we consider the same model [1.39] with:

$$B = \begin{pmatrix} 1 \\ x_1^2(t) \end{pmatrix}, \; C = \begin{pmatrix} x_1^2(t) & 0 \end{pmatrix}$$

then $h_1(x_2(t)) = \sin(x_2(t))$ and $h_2(x_1(t)) = x_1^2(t)$ represent nonconstant terms. Note that $h_1(x_2(t))$ is bounded for any $x(t)$, whereas $h_2(x_1(t))$ may be bounded for only a compact space $[-a,a]$, $a > 0$. In this case $h_1(x_2(t))$ and $h_2(x_1(t))$ can be written as follows:

$$h_1(x_2(t)) = F_{11}(x_2(t)).1 + F_{12}(x_2(t)).(-1)$$

$$h_2(x_1(t)) = F_{21}(x_1(t)).a^2 + F_{22}(x_1(t)).0$$

where:

$$F_{11}(x_2(t)) + F_{12}(x_2(t)) = 1$$

$$F_{21}(x_1(t)) + F_{22}(x_1(t)) = 1$$

$$F_{21}(x_1(t)) = \frac{x_1^2(t)}{a^2}; \quad F_{22}(x_1(t)) = 1 - \frac{x_1^2(t)}{a^2}.$$

$$F_{11}(x_2(t)) = \tfrac{1}{2}(1 + \sin(x_2(t))); \quad F_{12}(x_2(t)) = \tfrac{1}{2}(1 - \sin(x_2(t)))$$

Thus we get four local models, obtained from the four possible combinations of bounded terms $h_1(x_1(t))$ and $h_2(x_2(t))$, which are described by the following matrices A_i, B_i and C_i:

$$A_1 = \begin{pmatrix} -1 & 1 \\ 2 & -3 \end{pmatrix}, \quad B_1 = \begin{pmatrix} 1 \\ a^2 \end{pmatrix}, \quad C_1 = \begin{pmatrix} a^2 & 0 \end{pmatrix}$$

$$A_2 = \begin{pmatrix} -1 & -1 \\ 2 & -3 \end{pmatrix}, \quad B_2 = \begin{pmatrix} 1 \\ a^2 \end{pmatrix}, \quad C_2 = \begin{pmatrix} a^2 & 0 \end{pmatrix}$$

$$A_3 = \begin{pmatrix} -1 & 1 \\ 2 & -3 \end{pmatrix}, \quad B_3 = \begin{pmatrix} 1 \\ 0 \end{pmatrix}, \quad C_3 = \begin{pmatrix} 0 & 0 \end{pmatrix}$$

$$A_4 = \begin{pmatrix} -1 & -1 \\ 2 & -3 \end{pmatrix}, \quad B_4 = \begin{pmatrix} 1 \\ 0 \end{pmatrix}, \quad C_4 = \begin{pmatrix} 0 & 0 \end{pmatrix}$$

The corresponding activation functions are obtained from products $\boldsymbol{F}_{1(1or2)} \cdot \boldsymbol{F}_{2(1or2)}$:

$$\mu_1(x(t)) = F_{11}(x_1(t)).F_{21}(x_2(t))$$

$$\mu_2(x(t)) = F_{11}(x_1(t)).F_{22}(x_2(t))$$

$$\mu_3(x(t)) = F_{12}(x_1(t)).F_{21}(x_2(t))$$

$$\mu_4(x(t)) = F_{12}(x_1(t)).F_{22}(x_2(t))$$

Finally, the corresponding multiple model is given by:

$$\begin{cases} \dot{x}(t) = \sum_{i=1}^{4} \mu_i(x(t))(A_i x(t) + B_i u(t)) \\ y(t) = \sum_{i=1}^{4} \mu_i(x(t)) C_i x(t) \end{cases}$$

Note that the number of LTI models is dependent on the number of nonconstant terms contained in matrices A, B and C. In general, if s nonconstant terms are available, then the multiple model comprises at most 2^s local models.

In this example, in response to a unit step, the multiple model provides a good representation of the nonlinear system over the domain $U = [-a, a]$ with $a > 0$ (see Figure 1.9).

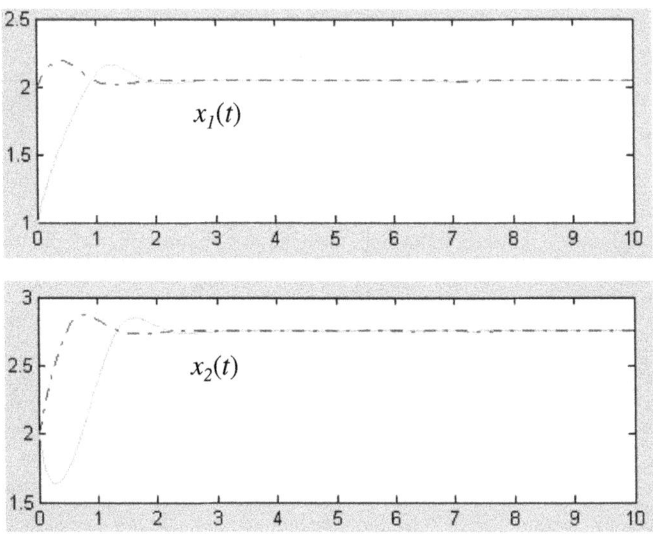

Figure 1.9. *Response of nonlinear model [1.39] and its corresponding multiple model*

EXAMPLE 1.4.– Let us consider the following nonlinear model:

$$\begin{cases} \dot{x}_1(t) = 5x_1(t) + \cos\left(x_1(t)\right)x_2\left(t\right) + u(t) \\ \dot{x}_2(t) = -x_1(t) + 2x_2(t) + x_2^3(t)u(t) \\ y(t) = x_2^3(t)x_1\left(t\right) + x_2(t) \end{cases} \qquad [1.40]$$

which can be rewritten in the form of [1.36], where:

$$A\left(x(t)\right) = \begin{pmatrix} 5 & \cos\left(x_1(t)\right) \\ -1 & 2 \end{pmatrix}, B\left(x(t)\right) = \begin{pmatrix} 1 \\ x_2^3(t) \end{pmatrix}$$

$$C\left(x(t)\right) = \begin{pmatrix} x_2^3(t) & 1 \end{pmatrix}, D\left(x(t)\right) = 0$$

The aim is to express $E\left(x(t)\right)$ defined in [1.37] in the form of [1.38]. Matrix $E\left(x(t)\right)$ presents $s = 2$ nonconstant terms:

$$h_1\left(x_1(t)\right) = \cos\left(x_1(t)\right), h_2\left(x_2(t)\right) = x_2^3(t) \qquad [1.41]$$

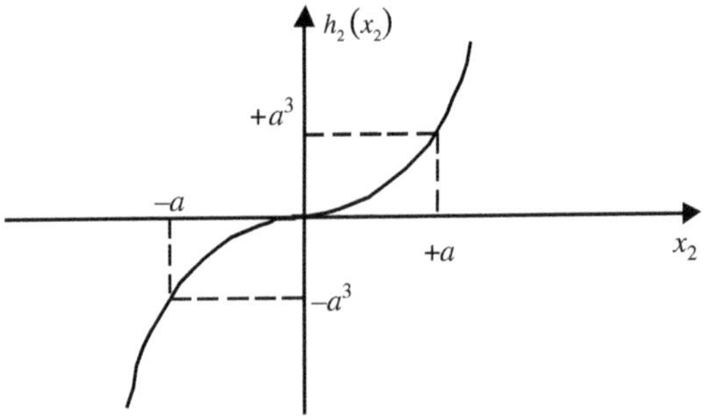

Figure 1.10. *Nonconstant term* $h_2\left(x_2\left(t\right)\right) = x_2^3\left(t\right)$

We observe that nonconstant term $h_1\left(x_1(t)\right)$ is bounded $\forall$ $x(t) \in \mathbb{R}^2$, whereas term $h_2\left(x_2(t)\right)$ can only be bounded over the bounded compact space $\left[-a, a\right]$, $a > 0$ (Figure 1.10).

Thus, nonlinear terms $h_1(x_1(t))$ and $h_2(x_2(t))$ $\forall x(t) \in U$, where $U = \mathbb{R} \times [-a, a], a > 0$, can be transformed such that:

$$h_1(x_1(t)) = F_1^1(x_1).1 + F_1^2(x_1).(-1) \tag{1.42a}$$

$$h_2(x_2(t)) = F_2^1(x_2).a^3 + F_2^2(x_2).(-a^3) \tag{1.42b}$$

where:

$$F_1^1(x_1) = \frac{1}{2}(1 + \cos(x_1(t))), F_1^2(x_1) = \frac{1}{2}(1 - \cos(x_1(t))) \tag{1.43a}$$

$$F_2^1(x_2) = \frac{1}{2a^3}(x_2^3(t) + a^3), F_2^2(x_2) = \frac{1}{2a^3}(a^3 - x_2^3(t)) \tag{1.43b}$$

Thus we obtain four LTI models represented by the matrices E_i described in [1.21]:

$$E_1 = \begin{pmatrix} 5 & 1 & 1 \\ -1 & 2 & a^3 \\ a^3 & 1 & 0 \end{pmatrix}, E_2 = \begin{pmatrix} 5 & 1 & 1 \\ -1 & 2 & -a^3 \\ -a^3 & 1 & 0 \end{pmatrix}$$

$$E_3 = \begin{pmatrix} 5 & -1 & 1 \\ -1 & 2 & a^3 \\ a^3 & 1 & 0 \end{pmatrix}, E_4 = \begin{pmatrix} 5 & -1 & 1 \\ -1 & 2 & -a^3 \\ -a^3 & 1 & 0 \end{pmatrix}$$

The activation functions are obtained from the products of $F_{1(1or2)} F_{2(1or2)}$:

$$\mu_1(x(t)) = F_1^1(x_1).F_2^1(x_2), \mu_2(x(t)) = F_1^1(x_1).F_2^2(x_2) \tag{1.44}$$

$$\mu_3(x(t)) = F_1^2(x_1).F_2^1(x_2), \mu_4(x(t)) = F_1^2(x_1).F_2^2(x_2) \tag{1.45}$$

The multiple model representation of a nonlinear system [1.40] is only valid over the state–space range $U = \mathbb{R} \times [-a, a]$. Consequently, the stability analysis of [1.39] based on its multiple model [1.1] is only local even if the stability established for the latter is global. For bounded matrices $E(x(t)) \quad \forall \quad x(t) \in \mathbb{R}^p$, the multiple model representation is confused with its nonlinear model. We recall that, as a general rule, if matrix $E(x(t))$ presents s nonconstant terms, then the multiple model consists of at most 2^s local models.

1.2.4. *Multiple model representation using the neural approach*

When there is no general starting model available, it often becomes difficult to create a multiple model representation of a process, and particularly to define *a priori* the base number of models to choose. In this respect, it is tempting to create a multiple model representation directly during the identification phase by using a neural approach [ELF 10a].

There are four steps to this approach:

– When input–output signals are considered to be sufficiently strong, to distribute this data into classes using a rival penalized competitive learning neural classification technique, this approach enables us to define the required number of models.

– To use the result of this classification in order to refine it from Kohonen self-adaptive networks and from the fuzzy K-means method.

– To associate a model, mostly linear, i.e. to each data set.

– To determine at each operating point the validity or coefficient relevant to each type of model.

1.2.4.1. *Determining the base number of models*

To determine these models, it is advisable, as for an identification, to excite the process using a frequency signal which is sufficiently strong and has varying amplitude, so as to excite all of the dynamics of the system and to obtain a good understanding of the nonlinear aspects.

This step in determining the optimum base number of models is particularly important; indeed if the number of models chosen is too low the representation would be unsatisfactory, whereas choosing an unreasonably high number would reduce the use of this type of representation.

Having collected these data, they will be grouped into classes using a rival penalized competitive learning neural network; this unsupervised learning strategy, proposed by [XU 93] and revised by [TAM 06], has recently been applied in [ELF 08, ELF 10b] in order to automatically determine multiple model representations. The main interest in this method lies in its ability to enable us to determine the number of data classes automatically [ELF 10b], and consequently the number of models to be implemented.

NOTE:

– $x(t)=\left[x_1(t),x_2(t),\ldots,x_n(t)\right]^T$ is the vector grouping the input–output data sets available at time t.

– $z_j^{(2)}(t)$ is the output from the jth neuron of the output layer at time t.

– $z^{(1)}(t)=\left\|w_j(t)-x(t)\right\|$.

– $w_j(t) = \left[w_{j1}(t), w_{j2}(t), \ldots, w_{jn}(t) \right]^T$ is the weight vector linking inputs from the *j*th neuron. This vector corresponds to class C_j.

– $z^{(2)}(t) = \arg\min_{j} \left\| w_j(t) - x(t) \right\|$.

From an arbitrarily chosen number of classes *K*, each one represented by a neuron, the data vectors are successively presented to the network. Let $x(t)$ be the data vector presented at time *t*.

NOTE:

– $g = \arg\min_{k} \left\| w_k(t) - x(t) \right\|$.

– $r = \arg\min_{k \neq g} \left\| w_k(t) - x(t) \right\|$.

The learning takes place by actualizing the weights as follows:

$$w_g(t+1) = w_g(t) + \eta(t)(x(t) - w_g(t))$$
$$w_r(t+1) = w_r(t) - \gamma(t)(x(t) - w_r(t))$$

the other weights being unchanged.

Having represented, multiple times if required, the various sets of data from the network, if learning coefficients $\eta(t)$ and $\gamma(t)$ are such that $\gamma(t)$ is much less than $\eta(t)$, then the network is able to determine the number of classes present.

Effectively:

– If the number of neurons *K* is chosen to be greater than the number of real classes present, the winning neurons evolve toward the center of the classes and the other centers diverge from the set of observations.

– If the number of output neurons K is less than the number of classes, there is oscillation of weight vectors between the various classes during the learning phase, which indicates a need to add one or more classes.

At the end of this first step, a first distribution of the data into classes takes place.

1.2.4.2. *More precise determination of the operating classes of the system*

This new step enables the distribution into classes to be refined in order to obtain various operating ranges from which the base models will be identified [BOR 06, BOR 07].

1.2.4.2.1. Classification using the Kohonen rule

This is similar to the previous rule by simply making $\gamma(t)=0$.

From the class centers previously obtained, the classification is brought about by applying the following algorithm for each instance of vector $x(t)$:

$$g=\arg\min_{k}\left\|w_k(t)-x(t)\right\|$$
$$w_g(t+1)=w_g(t)+\eta(t)(x(t)-w_g(t))$$
$$w_k(t+1)=w_k(t)\ \ \forall k\neq g$$

This operation is repeated until the weights are stabilized.

1.2.4.2.2. Classification using the fuzzy K-means algorithm

The fuzzy K-means algorithm enables us to group the elements using an approach which is fuzzy with a certain degree of membership, with each element having the ability to be part of multiple classes.

This algorithm is based on the minimization of the following objective function:

$$J_{FKM} = \sum_{i=1}^{N} \sum_{j=1}^{K} \mu_{ij}^{m} \left\| w_i(t) - x_j(t) \right\|^2$$

where:

m is the real number, $m \geq 1$

x_i is the vector corresponding to the ith data taken

μ_{ij} is the degree of membership of x_i to the class defined by w_j

N is the total number of observations

K is the number of classes previously determined

$\|v\|$ is the a norm of vector v

It is applied as follows:

– Initialize matrix $M = \{\mu_{ij}\}$ to value M_0, $M_k = \{\mu_{ij}^k\}$ value of M at the kth iteration.

– At step k, calculate vectors w_j which define the class centers:

$$w_j = \frac{\sum_{i=1}^{N} (\mu_{ij}^k)^m x_i}{\sum_{i=1}^{N} (\mu_{ij}^k)^m}$$

– Actualize M:

$$\mu^{k+1} = \{M_{ij}^{k+1}\}$$

$$\mu_{ij}^{k+1} = \left(\sum_{r=1}^{k} \left(\frac{\left\| x_i(t) - w_j(t) \right\|}{\left\| x_i(t) - w_r(t) \right\|} \right)^{\frac{2}{m-1}} \right)^{-1}$$

– If $\left\| M_{k+1} - M_k \right\| < \varepsilon$ (value fixed in advance to define the stop criterion) then end the algorithm, otherwise continue iteration.

Using this approach there is convergence toward a local minimum of J_{FKM}.

For the final classification, each vector x_i is attributed to the class for which it has the highest membership coefficient.

REMARK.– The first method presented enables the optimum number of classes to be determined. With regards to the precise determination of classes, experience shows that if the classes are well separated, the Kohonen rule gives the best result; on the other hand, if the classes overlap, then the best result is provided using the fuzzy K-means method.

1.2.4.3. *Identification of base models*

For each data class, it is first advisable to define the order of the model associated with it. The most satisfactory method appears to be the instrumental determinants ratio (IDR) test [BEN 01], which is based on the ranking conditions of the information matrix constructed from pairs of input–output measurements.

NOTE:

$$Q_m = \frac{1}{N_H} \sum_{k=1}^{N_H} \begin{pmatrix} u(k) \\ u(k+1) \\ \vdots \\ u(k+m) \end{pmatrix} \left(y(k+1),\, u(k+1), \cdots y(k+m),\, u(k+m) \right) \qquad [1.46]$$

where N_H is the number of observations and $u(k)$ and $y(k)$ are the control and output at time k, respectively.

The IDR is given by:

$$IDR(m) = \left| \frac{\det(Q_m)}{\det(Q_{m+1})} \right| \qquad [1.47]$$

The order of the model corresponds to the value of m for which a significant increase in the IDR(m) ratio is observed.

Identification is then carried out using classic methods such as the recursive least squares method.

It should be pointed out that, using this method, the base models are not necessarily of the same order.

1.2.4.4. *Calculation of the validity of models*

Between 0 and 1, validity is characteristic of the relevance of each model at a given instant [DEL 96].

A first very simple choice corresponds to the following determination.

NOTE:

$$r_i = |y - y_i| \quad i = 1, \ldots, N$$

where N is the number of base models, y the output from the system and y_i the output from the ith model:

$$r_i^{'} = \frac{r_i}{\sum_{i=1}^{N} r_i}, \; v_i^{'} = 1 - r_i^{'}, \; v_i^{simp} = \frac{v_i^{'}}{N-1} \qquad [1.48]$$

When the classes relative to the models are well separated, it is preferable to use reinforced validities [ELF 08]:

$$v_i'' = v_i^{simp} \prod_{i=1}^{N}\left(1 - v_i^{simp}\right),\ v_i^{renf} = \frac{v_i''}{\sum_{i=1}^{N} v_i''} \qquad [1.49]$$

1.3. Analysis and synthesis tools

1.3.1. *Lyapunov approach*

Lyapunov's direct method is based on the analysis of the total energy dissapted in a (mechanical) system: if this energy is continually dissipated (case of a dissipative system), then the system can be hoped to move toward an equilibrium point. Thus, the idea of the Lyapunov approach is to examine how a scalar function varies in order to study the stability of a given system [BOR 93, FOS 93, GRU 76, KAL 60].

Let us first consider the unforced nonlinear system described by:

$$\dot{x}(t) = f\left(x(t)\right) \qquad [1.50]$$

where $f\left(x(t)\right) \in C^1, f(.) : \mathbb{R}^n \rightarrow \mathbb{R}^n$

System [1.50] is said to be in equilibrium around x_0 if, in the absence of an external influence, its state does not vary over time; x_0 is thus called the equilibrium point.

DEFINITION 1.1 (EQUILIBRIUM POINT).– x_0 is said to be the equilibrium point of system [1.50] if $f(x_0) = 0, \forall t > t_0$, t_0 is the start time.

Later, we will consider, without loss of generality, that the origin of state space is the equilibrium point $(x_0 = 0)$ of system [1.50]. Indeed, if $x_0 \neq 0$ is the equilibrium point of [1.50], then $z_0 = 0$ is the equilibrium point of system $\dot{z} = f\left(z(t) + x_0\right)$.

THEOREM 1.1.– Let $V(x)$ be a scalar function which satisfies Lyapunov conditions; in particular, $V(x)$ is bounded by two positive scalar functions of $\|x(t)\|$, which are nondecreasing and continuously differentiable. If $\frac{\partial V(x)}{\partial x} f(x) < 0, \forall\, x(t) \neq 0$, then the equilibrium point $x(t) = 0$ of [1.50] is generally asymptotically stable.

Generally speaking, there is no method which enables a Lyapunov function to be chosen from all possible candidate functions. Therefore, Lyapunov's theory leads to sufficient stability conditions, the conservatism of which is dependent on the particular form chosen for function $V(x(t))$ and on the structure of the system. The most typical choice uses a quadratic form $V(x(t)) = x(t)^T P x(t)$, $P > 0$. This type of function, which has been adopted in order to study the stability of linear systems, is also used for multiple models [AKH 03, AKH 04, CHA 08, CHA 10, MA 98, MAR 99, MUR 97, TAN 98]. For the multiple model approach, convex formulation of the problem easily enables such a function to be extracted when it is possible. The disadvantage of the method is the obtaining of very conservative stability conditions. In order to reduce the conservatism of the quadratic stability, other types of Lyapunov functions are considered [CHA 00, CHA 06b, GUE 04, JAD 99, JOH 98, KRU 08, TAN 01]. Among these functions, the piecewise quadratic functions can be distinguished:

$$V\left(x(t)\right) = \max\left(V_1\left(x(t)\right), \ldots, V_i\left(x(t)\right), \ldots, V_n\left(x(t)\right)\right) \qquad [1.51]$$

where:

$$V_i\left(x(t)\right) = x(t)^T P_i x(t), P_i > 0, i \in I_N \qquad [1.52]$$

This function, which will be used later in this book, can be seen as a more general form than quadratic functions.

1.3.2. *Numeric tools: linear matrix inequalities*

The formulation in terms of convex optimization of analysis and synthesis problems is the fundamental reason for using the multiple model approach. There are two advantages relating to the convexity of an optimization problem: i) reasonable computation times in order to find a solution and ii) the result obtained corresponds to a unique global minimum.

Convexity is a concept which concerns both convex functions and convex sets, hence the following definitions:

– A set $E \subset \mathbb{R}^n$, where E is a convex set, if and only if:

$$\forall \lambda \in [0,1] \subset \mathbb{R}, \forall (x_1, x_2) \in E^2, \lambda x_1 + (1-\lambda) x_2 \in E \qquad [1.53]$$

– A function $f:\ E \subset \mathbb{R}^n \rightarrow \mathbb{R}$, where E is a convex set, then f is convex if and only if:

$$\forall \lambda \in [0,\ 1] \subset \mathbb{R}, \forall (x_1, x_2) \in E^2, f\left(\lambda x_1 + (1-\lambda) x_2\right) \leq \lambda f(x_1) + (1-\lambda) f(x_2) \qquad [1.54]$$

A convex optimization problem is therefore set out as follows: $\min_{x \in E}\left(f(x)\right)$, where E is a convex set and f a convex function.

Many automation problems are unfortunately nonconvex. For this type of problem there are methods which make conditions convex. However, we must bear in mind that transforming a problem from nonconvex into a convex formulation may lead to reduced validity domains and conservative constraints. Thus, although the solution to the new problem obtained through convex optimization may be optimal, it is not necessarily optimal in the sense of the initial problem. The obtained solution is also optimal for the initial problem in only the case where convex/nonconvex transformation occurs through equivalence. This remark is fundamental for interpreting the results obtained using optimization methods. In the next section we recall some of the main results of this formalism.

1.3.2.1. *Linear matrix inequality formalism*

For a number of years, numerous studies within automation have been formulated in terms of convex optimization, implying linear matrix inequalities (LMI). Reasons for the popularity of this approach include the existence of effective numeric resolution methods and the possibility of handling multicriteria synthesis. The following are some basic ideas for the LMI approach [BOY 94, CHI 96, RIC 01].

DEFINITION 1.2.– Let P_0 and $P_i, i \in I_n$ be a family of symmetric matrices of $\mathbb{R}^{p.p}$ and vector $x = (x_1, x_2, ..., x_n)^T \in \mathbb{R}^n$, a strict (or nonstrict, respectively) LMI in $x_i, i \in I_n$ is expressed in the form:

$$F(x) = P_0 + \sum_{i=1}^{n} x_i P_i > 0 \, (resp. \geq 0) \quad [1.55]$$

Note that the set E defined by $E = \{x \in \mathbb{R}^n : F(x) > 0\}$ is convex.

The three most common convex optimization problems in LMI form are:

– The problems of feasibility, i.e. where a solution is accepted): the issue here is to find a vector x such that the convex constraint $F(x) > 0$ can be satisfied. This type of problem can be resolved by finding the vector x which minimizes the scalar t such that:

$$-F(x) < t.\mathcal{I} \qquad [1.56]$$

If the minimum value of t is negative, the problem is said to be feasible

– Eigenvalue problems: the largest eigenvalue of a symmetric matrix A must be minimized, under an LMI constraint which is linear in B:

- minimize λ

- under constraints

$$\begin{cases} \lambda\mathcal{I} - A > 0 \\ B > 0 \end{cases} \qquad [1.57]$$

These convex optimization problems can thus be resolved using various types of methods [BOY 94, RIC 01].

1.3.2.2. *Analysis elements*

Some of the most used tools for solving LMI constraints are presented in the following.

1.3.2.2.1. Schur complement

LEMMA 1.1.– Let $R = R^T$, $Q = Q^T$ and S be three matrices. The following LMI are equivalent:

$$\text{i)} \begin{pmatrix} Q & S \\ S^T & R \end{pmatrix} > 0 \qquad [1.58]$$

$$\text{ii)}\ R > 0,\ Q - SR^{-1}S^T > 0 \qquad [1.59]$$

$$\text{iii)}\ Q > 0,\ R - S^T Q^{-1} S > 0 \qquad [1.60]$$

1.3.2.2.2. Convex quadratic constraints

The constraint over norm $\|Z\| < 1$, where $Z \in \mathbb{R}^{p.q}$, is represented by:

$$\begin{pmatrix} \mathcal{I}_p & Z \\ Z^T & \mathcal{I}_q \end{pmatrix} > 0 \qquad [1.61]$$

Another technique, known as the S-procedure, is often used in synthesis using an LMI approach. This is a technique which enables a set of convex or nonconvex constraints to be approached using a convex constraint.

LEMMA 1.2.– Let $F_0,\ldots,F_q$ be quadratic functions dependent on variables $\xi \in \mathbb{R}^p$, such that $F_i(\xi) = \xi^T T_i \xi + 2u_i \xi + v_i, i \in I_q$ and $T_i = T_i^T \in \mathbb{R}^{p.p}$. Thus, proposition i) implies proposition ii).

i) $\exists\ \tau_0 \geq 0,\ldots,\tau_q \geq 0$, such that $F_0(\xi) - \sum_{i=1}^{q} \tau_i F_i(\xi) \leq 0$

ii) $F_0(\xi) \leq 0\ \ \forall \xi \neq 0$, such that $F_i(\xi) \leq 0, i \in I_q$

If $q = 1$, then the reciprocal is also true.

1.3.2.2.3. Projection lemma

The following lemma, which is also called the elimination lemma, is one of the tools used in controller synthesis.

LEMMA 1.3.– For real matrices $W = W^T$, M, N of appropriate dimensions, the following properties are equivalent:

– There is a real matrix K such that:

$$W + MKN^T + NK^T M^T < 0 \qquad [1.62]$$

– There is a scalar σ such that:

$$W < \sigma MM^T, W < \sigma NN^T \qquad [1.63]$$

– Orthogonal complements $M^\perp$ and $N^\perp$ of M and N, respectively, verify:

$$M^{\perp T} W M^\perp < 0, \ N^{\perp T} W N^\perp < 0 \qquad [1.64]$$

The Schur complement, like the elimination lemma, is widely used in linearization of design conditions for controllers and observers, as we will see later in this book. The following discrete-time linear example illustrates these techniques:

$$x(k+1) = Ax(k) + Bu(k) \qquad [1.65]$$

The stability of the closed-loop via control law, $u(k) = Kx(k)$, consists of verifying the matrix inequality:

$$(A+BK)^T P(A+BK) - P < 0, \ P = P^T > 0 \qquad [1.66]$$

As we can see, this matrix inequality is nonlinear in K and P and can be linearized as follows:

– By multiplying both sides by $X = P^{-1} > 0$, the following equivalent inequality is obtained:

$$X(A+BK)^T X^{-1}(A+BK)X - X < 0, \ X = X^T > 0 \qquad [1.67]$$

– By applying the Schur complement, the following equivalent inequality is obtained:

$$\begin{pmatrix} X & X(A+BK)^T \\ * & X \end{pmatrix} > 0 \qquad [1.68]$$

– By applying the change in variable $Y = KX$, the following LMI, which is equivalent to the initial problem [1.66] is obtained:

$$\begin{pmatrix} X & (AX+BY)^T \\ * & X \end{pmatrix} > 0 \qquad [1.69]$$

1.3.2.3. *Bilinear matrix inequalities*

Many design problems (such as control and observation) are expressed in bilinear matrix inequality (BMI) form and are difficult to solve using existing numeric tools developed for LMI problems. Currently, there is no exact method for solving these types of problem, but some heuristics have been developed.

DEFINITION 1.3.– A BMI in $x_i, i \in I_n$, is a constraint of the form:

$$F(x) = P_0 + \sum_{i=1}^{n} x_i P_i + \sum_{i=1}^{n}\sum_{j=1}^{n} x_i x_j P_{ij} > 0 \ \ (\text{resp.} \geq 0) \qquad [1.70]$$

where matrices P_i and P_{ij} belonging to $\mathbb{R}^{n.n}$ are symmetric.

Problems formulated in BMI terms may have multiple local solutions and cannot be resolved using the convex optimization techniques developed for solving LMI problems, as they are often difficult nonconvex problems. Some heuristic methods which enable solutions to these types of problems have been proposed. However, methods exist which enable some BMI problems to be reduced to LMI problems.

For example, let us consider the following BMI problem which is dependent on variables K_i and $P > 0$:

$$\left(A_i + B_i K_i\right)^T P + P\left(A_i + B_i K_i\right) < 0, \forall i \in I_N \qquad [1.71]$$

We can solve this type of problem using two equivalent linearization techniques:

– By a change in variables: the product of two variables, which appears in the synthesis conditions, is replaced by a new variable. Indeed, by pre- and post-multiplying the inequalities [1.71] by $P^{-1} = Q$ we obtain:

$$Q > 0,\ Q\left(A_i + B_i K_i\right)^T + \left(A_i + B_i K_i\right) Q < 0 \qquad [1.72]$$

By inserting the change in variables, $Y_i = K_i Q$, the following LMI in Y_i and Q are obtained:

$$Q > 0,\ A_i Q + Q A_i^T + B_i Y_i + Y_i^T B_i^T < 0 \qquad [1.73]$$

This is an LMI problem equivalent to a BMI problem [1.71];

– By using the projection lemma. Indeed, by pre- and post-multiplying inequalities [1.73] by $B_i^{\perp}$, the orthogonal complement of B_i, the following equivalent LMI are obtained:

$$Q > 0,\ B_i^{\perp T}\left(A_i Q + Q A_i^T\right) B_i^{\perp} < 0 \qquad [1.74]$$

Once variables K_i are eliminated and matrix Q is calculated from [1.74], the variables K_i are obtained by solving the LMI problems [1.71].

1.3.3. *Multiple model control techniques*

In order to develop control laws for multiple models, various techniques have been adapted and introduced.

1.3.3.1. *Parallel distributed compensation control law*

The parallel distributed compensation (PDC) method [TAN 98], which was developed for multiple models, is constructed by interpolation, through the same controller activation functions, by full state feedback relative to each local LTI model. It is expressed in the form:

$$u(t)=\sum_{i=1}^{N}\mu_i\left(z(t)\right)K_i x(t) \qquad [1.75]$$

A particular example of this control law is the linear full state feedback of the form $u(t)=Kx(t)$. However, this method is very restrictive since it is easy to find some examples where no stabilizing linear control law exists, whereas a PDC control law exists. This assessment is justified by the fact that common variable K needs to stabilize n local models, whereas for the PDC controller, we consider the recovery rate of the activation functions.

1.3.3.2. *Other control laws*

Other control laws, which are slightly different from the PDC control, have also been studied:

– The first control law of the form:

$$u(t)=\frac{\sum_{i=1}^{N}\mu_i\left(z(t)\right)\alpha_i K_i}{\sum_{i=1}^{N}\mu_i\left(z(t)\right)\alpha_i}x(t) \qquad [1.76]$$

is interesting for positively linearly dependent input matrices (B_i), i.e. $\exists\alpha_i>0: B_i=\alpha_i B, \forall\ i\in I_N$ [CHA 02b, GUE 04]. The closed loop is expressed without crossed models and the stabilization conditions are those of stability (by substituting A_i with $A_i - B_i K_i$), which allows us to reduce the number of constraints arising from this control law to N LMI instead of $N(N+1)/2$ for the PDC control law.

– The second control law is of the form:

$$\dot{u}(t) = A_c u(t) + \sum_{i=1}^{N} \mu_i \left(z(t) \right) K_i x(t) \qquad [1.77]$$

where $A_c \in \mathbb{R}^{n.n}$ and $K_i \in \mathbb{R}^{m.n}$. This is the PDC control law, filtered using a low pass filter $G(s) = (sI - A_c)^{-1}$.

Other control laws derived from the PDC control law have been used, including the *Proportional PDC* (PPDC) control law, or the PDC control law increased from the term $\sum_{i=1}^{N} \frac{\partial \mu_i \left(z(t) \right)}{\partial t} K_i x(t)$ [CHA 02b].

1.3.3.3. *Control laws based on output feedback*

Output PDC (OPDC) controller is of the form:

$$u(t) = \sum_{i=1}^{N} \mu_i \left(z(t) \right) F_i y(t) \qquad [1.78]$$

This is a static control law based on output feedback. In the particular example where input matrices B_i are linearly dependent, the control law above is modified by basing it on the PDC control law [CHA 02b]:

$$u(t)=\frac{\sum_{i=1}^{N}\mu_i\left(z(t)\right)\alpha_i F_i}{\sum_{i=1}^{N}\mu_i\left(z(t)\right)\alpha_i}y(t) \qquad [1.79]$$

This control law is continuous and verifies $\sum_{i=1}^{N}\mu_i\left(z(t)\right)\alpha_i \geq \min_{i\in I_n}(\alpha_i) > 0$.

Other nonlinear dynamic control laws based on output feedback have also been studied [HA 03]. The resulting stabilization conditions are in BMI form. Some linearization algorithms have been proposed [BOY 94, HAS 99].

Chapter 2

Stability of Continuous Multiple Models

2.1. Introduction

Our objective in this chapter is to study the stability of multiple models. Numerous studies on the subject have initially been inspired from techniques developed in the linear domain. Indeed, Lyapunov's method, or more specifically the quadratic method, and formulation using linear matrix inequalities have been extensively used [BOY 94, CHA 02b, CHA 08d, CHA 10, MA 98, MAR 99, TAN 98]. In this sense, transformation into a Lur'e problem, interconnected system techniques and the properties of M-matrices have been adapted for nominal and uncertain multiple models [BRE 02, CHA 02b].

This chapter deals with the stability of continuous multiple models, with and without uncertainties. Two types of Lyapunov function are used, quadratic and nonquadratic, and the chapter is set out as follows. In the following section, we propose some sufficient conditions for standard multiple model stability, based on the presence of quadratic Lyapunov functions and the properties of M-matrices. Then, we present some conditions for stability which make use of nonquadratic functions. Finally, we examine conditions for robust stability, which imply two types of uncertainty.

2.2. Stability analysis

The approach proposed in this chapter is principally based on quadratic Lyapunov functions. We need to find a positive definite symmetric matrix, i.e. the associated quadratic Lyapunov function which guarantees asymptotic multiple model stability.

Let us consider the continuous multiple model of the form:

$$\dot{x}(t) = \sum_{i=1}^{N} \mu_i\left(z(t)\right)\left(A_i x(t) + B_i u(t)\right) \qquad [2.1]$$

where:

$$\mu_i\left(z(t)\right) \geq 0, \sum_{i=1}^{N} \mu_i\left(z(t)\right) = 1 \qquad [2.2]$$

The open-loop multiple model which corresponds to [2.1] is:

$$\dot{x}(t) = \sum_{i=1}^{N} \mu_i\left(z(t)\right) A_i x(t) \qquad [2.3]$$

Multiple model [2.3] is globally asymptotically stable if a symmetric matrix $P > 0$ exists such that the following LMI are verified:

$$P > 0, A_i^T P + PA_i < 0, \forall i \in I_N \qquad [2.4]$$

This result is obtained by deriving, for the whole trajectory of multiple model [2.3], the quadratic function $V(x) = x^T(t)Px(t)$. The existence of $P > 0$ is dependent on two conditions:

– The first is linked to the stability of all local models. Each matrix A_i must be a Hurwitz matrix[1].

– The second relates to the existence of a Lyapunov function which is common to the N local models. This requires matrix $\sum_{i=1}^{N} A_i$ to be a Hurwitz matrix. Proof may be obtained by summing LMI [2.4]. This condition provides us with a quick test. Indeed, it is enough for two matrices to exist such that the sum $A_i + A_j$ is not stable in order to conclude the nonexistence of a symmetric matrix $P > 0$ satisfying the LMI [2.4]. The following lemma provides another means of carrying out this test.

LEMMA 2.1.– The following two propositions are equivalent [BOY 94]:

– Matrices Q_i, not all null, exist such that:

$$Q_1 \geq 0, \ldots, Q_N \geq 0, \sum_{i=1}^{N} \left(Q_i A_i^T + A_i Q_i \right) \geq 0 \qquad [2.5]$$

– No matrix P verifying the LMI [2.4] exists.

EXAMPLE 2.1.–

1) Let us consider a two-local-model multiple model defined by:

$$A_1 = \begin{pmatrix} -1 & 4 \\ 0 & -2 \end{pmatrix} \text{ and } A_2 = \begin{pmatrix} -1 & 0 \\ 4 & -2 \end{pmatrix}$$

which are two asymptotically stable matrices. Since matrix $A_1 + A_2 = \begin{pmatrix} -2 & 4 \\ 4 & -4 \end{pmatrix}$ is eigenvalue unstable, (1,12) and (–7,12),

1 A matrix for which the eigenvalues belong to the left half-plane of the complex plane.

then there is no common positive definite matrix which verifies [2.4].

2) The following is a multiple model consisting of two LTI models:

$$A_1 = \begin{pmatrix} -5 & -4 \\ -1 & -2 \end{pmatrix} \text{ and } A_2 = \begin{pmatrix} -2 & -4 \\ 20 & -2 \end{pmatrix}$$

which are two asymptotically stable matrices, as is their sum. However, there is no common positive definite matrix which verifies [2.4], although constraint [2.5] is verified with:

$$Q_1 = \begin{pmatrix} 90.0146 & -182.3584 \\ -182.3584 & 403.4566 \end{pmatrix} \text{ and } Q_2 = \begin{pmatrix} 44.2614 & 40.9912 \\ 40.9912 & 74.0205 \end{pmatrix}$$

Another way to study the stability of multiple models is to consider the properties of M-matrices and Lyapunov candidate scalar functions of the form:

$$V(x(t)) = x(t)^T \sum_{i=1}^{N} \alpha_i P_i x(t),\ P_i > 0,\ \alpha_i \in \mathbb{R}^{*+} \qquad [2.6]$$

This is a radially unbounded function:

$$c_1 \|x(t)\|^2 \leq V(x(t)) \leq c_2 \|x(t)\|^2 \qquad [2.7a]$$

where $c_1 = \lambda_{\min}\left(\sum_{i=1}^{N} \alpha_i P_i\right) > 0$ and $c_2 = \lambda_{\max}\left(\sum_{i=1}^{N} \alpha_i P_i\right) > 0$.

THEOREM 2.1.– Let us assume that symmetric matrices $P_i > 0,\ S_{ii} > 0$ and S_{ij} exist such that:

1) $A_i^T P_i + P_i A_i \leq -S_{ii},\ \forall\ i \in I_N$ [2.7b]

2) $A_j^T P_i + P_i A_j \leq S_{ji}, \ \forall \ (i,j) \in I_N^2, i \neq j$ [2.7c]

3) Φ is an M-matrix [2.7d]

where:

$$\Phi = \begin{pmatrix} \lambda_{\min}(S_{11}) & -|\lambda_{\max}(S_{12})| & \cdots & -|\lambda_{\max}(S_{1N})| \\ -|\lambda_{\max}(S_{21})| & \lambda_{\min}(S_{22}) & \cdots & \vdots \\ \vdots & \vdots & \ddots & -|\lambda_{\max}(S_{(N-1)N})| \\ -|\lambda_{\max}(S_{N1})| & \cdots & -|\lambda_{\max}(S_{N(N-1)})| & \lambda_{\min}(S_{NN}) \end{pmatrix}$$ [2.7e]

then multiple model [2.1] is globally asymptotically stable.

PROOF.– Let us consider Lyapunov function [2.6]. Its derivative along multiple model [2.1] gives:

$$\begin{aligned} \dot{V}(x(t)) &= x(t)^T \sum_{i=1}^{N}\sum_{j=1}^{N} \alpha_i \mu_j(z(t))\left(A_j^T P_i + P_i A_j\right)x(t) \\ &= x(t)^T \left(\sum_{i=1}^{N} \alpha_i \mu_i(z(t))\left(A_i^T P_i + P_i A_i\right) + \sum_{j \neq i:1}^{N} \alpha_i \mu_j(z(t))\left(A_j^T P_i + P_i A_j\right) \right) x(t) \end{aligned}$$ [2.8]

with conditions 1) and 2) of Theorem 2.1, equality [2.8] becomes:

$$\dot{V}(x(t)) \leq x(t)^T \left(-\sum_{i=1}^{N} \alpha_i \mu_i(z(t)) S_{ii} + \sum_{j \neq i:1}^{N} \alpha_i \mu_j(z(t)) S_{ij} \right) x(t)$$ [2.9]

We know that:

$$\forall x(t): \left|\lambda_{\min}\left(S_{ij}\right)\right| \left\|x(t)\right\|^2 \leq \left|x(t)^T S_{ij} x(t)\right| \leq \left|\lambda_{\max}\left(S_{ij}\right)\right| \left\|x(t)\right\|^2 \qquad [2.10]$$

With the definitions:

$$\begin{aligned} &\mu(z(t)) = \begin{pmatrix}\mu_1(z(t)) & \mu_2(z(t)) & \cdots & \mu_N(z(t))\end{pmatrix} \\ &\alpha = \begin{pmatrix}\alpha_1 & \alpha_2 & \cdots & \alpha_N\end{pmatrix}^T \end{aligned} \qquad [2.11]$$

inequality [2.9] leads to:

$$\begin{aligned} \dot{V}(x(t)) &\leq -\sum_{i=1}^{N} \alpha_i \mu_i(z(t)) \lambda_{\min}(S_{ii}) \left\|x(t)\right\|^2 \\ &\quad + \sum_{i \neq j:1}^{N} \alpha_i \mu_j(z(t)) \left|\lambda_{\max}(S_{ij})\right| \left\|x(t)\right\|^2 \\ &= -\mu(z(t)) \Phi \alpha \left\|x(t)\right\|^2 \end{aligned} \qquad [2.12]$$

where Φ is as defined in [2.7a].

Finally, as Φ is an M-matrix (see Appendix 2), then a positive vector exists such that vector $\beta = \Phi\alpha$ is positive. Consequently, $\dot{V}(x(t)) < 0$.

The problem with this method is that it is not easy to obtain an M-matrix. However, the following heuristic may be used:

1) In [2.7a], substitute matrix S_{ii} with $\overline{S}_{ii} = S_{ii} + \tau\mathcal{I}$, where τ is a real parameter and $\mathcal{I}$ is the appropriately sized identity matrix.

2) By iteration over τ, find an M-matrix [2.7e] which verifies conditions [2.7b]–[2.7d].

The following example illustrates this heuristic [CHA 02b, CHA 08b].

EXAMPLE 2.2.– Let us consider the multiple model described by the following two local models:

$$A_1 = \begin{pmatrix} 0 & 1 \\ -0,06 & -1 \end{pmatrix}, A_2 = \begin{pmatrix} 0 & 2,04 \\ -0,5 & -1,5 \end{pmatrix}$$

Direct application of the conditions of Theorem 2.1 does not lead us to an M-matrix. Consequently, nothing can be concluded as regards the stability of the system. However, by using the above-mentioned heuristic, the following LMI system is obtained:

$$P_1 > 0, P_2 > 0, S_{11} + \tau I > 0, S_{22} + \tau I > 0$$

$$A_1^T P_1 + P_1 A_1 \leq -\left(S_{11} + \tau I\right)$$

$$A_2^T P_2 + P_2 A_2 \leq -\left(S_{22} + \tau I\right)$$

$$A_1^T P_2 + P_2 A_1 \leq S_{12}$$

$$A_2^T P_1 + P_1 A_2 \leq S_{21}$$

By iteration over τ, we obtain:

$$P_1 = \begin{pmatrix} 1.0944 & 1.0115 \\ 1.0115 & 1.6493 \end{pmatrix}, P_2 = \begin{pmatrix} 0.8462 & 0.6593 \\ 0.6593 & 1.3542 \end{pmatrix}$$

$$\tau = 0,3, \; S_{11} = \begin{pmatrix} 60.5371 & 7.9986 \\ 7.9986 & 637.6728 \end{pmatrix}, \; S_{22} = \begin{pmatrix} 329.5031 & -30.1390 \\ -30.1390 & 686.1743 \end{pmatrix}$$

$$S_{12} = \begin{pmatrix} 666.8745 & 105.6867 \\ 105.6867 & -643.8083 \end{pmatrix}, S_{21} = \begin{pmatrix} -265.4610 & -109.2164 \\ -109.2164 & -75.1089 \end{pmatrix}$$

and

$$\Phi = \begin{pmatrix} 60.7263 & -675.3418 \\ -25.4171 & 327.2742 \end{pmatrix}$$

Since we are dealing with an M-matrix, it can be concluded that multiple model [2.26] is globally asymptotically stable, which can be shown using the Lyapunov function:

$$V(x(t)) = x(t)^T (\alpha_1 P_1 + \alpha_2 P_2) x(t)$$

where, for example $\alpha_2 = 1$ and $\alpha_1 \in [11,12 \quad 12,87]$.

2.2.1. *Exponential stability*

Exponential stability may be considered for the results above. This can be studied by considering the exponential system state decay rate defined by the largest scalar $\alpha > 0$ such that $\lim_{t \to \infty} e^{\alpha t} \|x(t)\| = 0$.

In order to prove the exponential stability of multiple model [2.1], the stability of the following system must be proven:

$$\dot{x}(t) = \sum_{i=1}^{N} \mu_i(z(t))(A_i + \alpha I)x(t) \qquad [2.13]$$

In order to study the stability of the initial system, it is equally valid to take the Lyapunov function:

$$W(x(t)) = V(x(t))e^{2\alpha t}$$

where:

$$V(x(t)) = x(t)^T Px(t), \ P > 0$$

The α-positive exponential stability is assured if $\forall\, x(t) \neq 0$:

$$\exists P > 0 \text{ such that } \frac{dV(x(t))}{dt} + 2\alpha V(x(t)) < 0 \qquad [2.14]$$

i.e.:

$$A_i^T P + PA_i + 2\alpha P < 0, \forall i \in I_N$$

The system is then said to be α-stable, where α is a minorant of the trajectory decay rate of system [2.1], which is important to maximize. Thus, maximizing the decay rate minorant is a generalized eigenvalue problem (GEVP) in P and α [BOY 94]. Its formulation is as follows:

$$\begin{aligned} &\textit{Maximize} \quad \alpha \\ &\textit{Objective} \ \ P > 0 \\ &\qquad A_i^T P + PA_i + 2\alpha P < 0, i \in I_N \end{aligned} \qquad [2.15]$$

2.3. Relaxed stability

Quadratic stability often introduces much conservatism (omission of activation functions and of their supports). Studies enabling conservatism from the quadratic method to be reduced have been proposed [CHA 00, CHA 06b, GUE 04, JAD 99, JOH 98, KRU 08, TAN 01]. In this context, we should also mention the techniques which use vector norms based on the Borne–Gentina criterion [BOR 74, GEN 72]. In order to relax the results above, the following Lyapunov function can be considered [CHA 06b]:

$$V\left(x(t)\right)=\max\left(V_1(x),...,V_i(x),...,V_N(x)\right) \quad [2.16]$$

where:

$$V_i\left(x(t)\right)=x(t)^T P_i x(t), P_i>0, \quad i\in I_N \quad [2.17]$$

These are Lyapunov candidate functions, particularly since $V\left(x(t)\right)>0, \forall \; x(t)\neq 0$ and $V(0)=0$.

The following theorem gives satisfactory conditions for multiple model stability:

$$\dot{x}(t)=\sum_{i=1}^{N}\mu_i\left(z(t)\right)A_i x(t) \quad [2.18]$$

by using the S-procedure (see section 1.3.2.2) and Lyapunov candidate function [2.16].

THEOREM 2.2.– Let us assume that matrices $P_i>0$ and scalars $\tau_k^{ij}\geq 0$ exist such that:

$$A_i^T P_j + P_j A_i + \sum_{k=1}^{N}\tau^{ij}{}_k\left(P_j - P_k\right)<0 \quad \forall(i,j)\in I_N^2 \quad [2.19]$$

then multiple model [2.18] is asymptotically stable.

PROOF.– Let us consider Lyapunov candidate function [2.16]. From this:

$$V\left(x(t)\right)=V_j\left(x(t)\right) \text{ if } V_j\left(x(t)\right)\geq V_k\left(x(t)\right) \; \forall\; (j,k)\in I_N^2, k\neq j \quad [2.20]$$

Indeed, in this example we have:

$$\frac{dV\left(x(t)\right)}{dt}=\frac{dV_j\left(x(t)\right)}{dt} \quad [2.21]$$

The following is obtained:

$$\frac{dV\left(x(t)\right)}{dt}=x(t)^T\sum_{i=1}^{N}\mu_i\left(z(t)\right)\left(A_i^T P_j+P_j A_i\right)x(t) \quad [2.22]$$

Thus, multiple model [2.18] is asymptotically stable if the following conditions are verified $\forall\ (i,j,k)\in I_N^3$:

$$x(t)^T\left(A_i^T P_j+P_j A_i\right)x(t)<0 \text{ for } x(t)^T\left(P_j-P_k\right)x(t)\geq 0 \ \ \forall x(t)\neq 0 \quad [2.23]$$

Conditions [2.19] are obtained through direct application of the S-procedure (see section 1.3.2.2) to conditions [2.23].

Note that the stability conditions arising from the quadratic approach are included in conditions [2.19]. Thus, when $P_i=P, i\in I_N$, function [2.16] becomes $V\left(x(t)\right)=x(t)^T Px(t)$, $\tau_k^{ij}\left(P_j-P_k\right)=0$ and conditions [2.19] are reduced to $A_i^T P+PA_i<0$. It should also be pointed out that these conditions are bilinear in τ_k^{ij} and P_i. A linearization heuristic could be used by fixing scalars τ_k^{ij}, and constraints [2.19] become LMI in P_i:

1) Fix the positive scalars τ_k^{ij} in [2.19].

2) Solve LMI [2.19] in P_i. If a solution exists then STOP. Otherwise return to 1) and modify τ_k^{ij}.

For example, by way of illustration for a multiple model with two-local-model (N = 2), the conditions to be verified are:

$$A_1^T P_2 + P_2 A_1 + \tau_1^{12}(P_2 - P_1) < 0 \text{ and } A_2^T P_1 + P_1 A_2 + \tau_2^{21}(P_1 - P_2) < 0$$

In certain cases, a simplified form of conditions [2.19] may be sufficient:

$$A_1^T P_2 + P_2 A_1 + \tau(P_2 - P_1) < 0 \text{ and } A_2^T P_1 + P_1 A_2 + \tau(P_1 - P_2) < 0$$

Note that this form is more conservative than the initial form [2.19].

2.4. Example

Let us consider the multiple model [2.18] with N = 3 such that:

$$A_1 = \begin{pmatrix} 0 & 1 \\ -0.06 & -1 \end{pmatrix}, \; A_2 = \begin{pmatrix} 0 & 1 \\ -1.94 & -1 \end{pmatrix}, \; A_3 = \begin{pmatrix} 0 & 2.04 \\ -0.5 & -1.5 \end{pmatrix} \quad [2.24]$$

The activation functions (Figure 2.1) are defined by:

$$\mu_1(x_1(t)) = \frac{\omega_1(x_1(t))}{\sum_{i=1}^{3} \omega_i(x_1(t))},$$

$$\mu_2(x_1(t)) = \frac{\omega_2(x_1(t))}{\sum_{i=1}^{3} \omega_i(x_1(t))},$$

$$\mu_3(x_1(t)) = \frac{\omega_3(x_1(t))}{\sum_{i=1}^{3} \omega_i(x_1(t))} \quad [2.25]$$

where:

$$\omega_1(x_1(t)) = \exp\left(-\frac{1}{2}\left(\frac{x_1(t)+5}{2}\right)^2\right), \omega_2(x_1(t)) = \exp\left(-\frac{1}{2}\left(\frac{x_1(t)}{2}\right)^2\right),$$

$$\omega_3(x_1(t)) = \exp\left(-\frac{1}{2}\left(\frac{x_1(t)-5}{2}\right)^2\right) \quad [2.26]$$

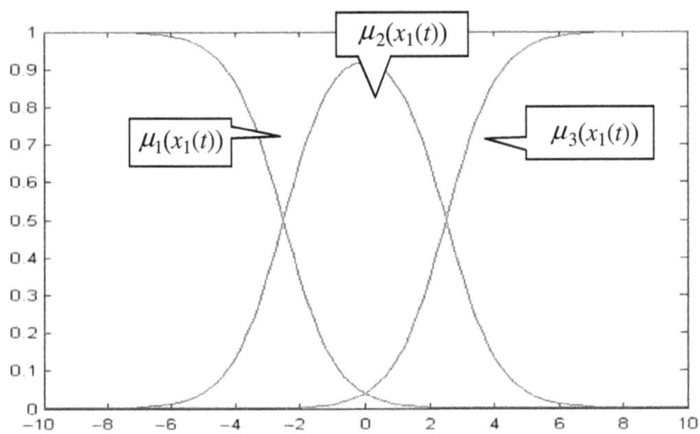

Figure 2.1. *Activation functions*

The quadratic method fails to prove the stability of this model. This can easily be verified by considering Lemma 2.1, i.e. that matrices $Q_1 > 0, Q_2 > 0, Q_3 > 0$ exist, such that the following conditions are verified:

$$Q_1 A_1^T + A_1 Q_1 + Q_2 A_2^T + A_2 Q_2 + Q_3 A_3^T + A_3 Q_3 > 0$$

The stability conditions [2.19] to be solved are:

$$A_i^T P_1 + P_1 A_i + \sum_{k=2}^{3} \tau_{i1k} (P_1 - P_k) < 0, i \in I_3$$

$$A_i^T P_2 + P_2 A_i + \sum_{k=1,k\neq 2}^{3} \tau_{i2k}\left(P_2 - P_k\right) < 0, i \in I_3$$

$$A_i^T P_3 + P_3 A_i + \sum_{k=1}^{2} \tau_{i3k}\left(P_3 - P_k\right) < 0, i \in I_3$$

Resolution of these conditions leads to:

$$\tau_2^{11} = \tau_3^{11} = 0, \tau_2^{21} = \tau_3^{21} = 1, \tau_2^{31} = \tau_3^{31} = 0$$

$$\tau_1^{12} = \tau_3^{12} = 0, \tau_1^{22} = \tau_3^{22} = 1, \tau_1^{32} = \tau_3^{32} = 0$$

$$\tau_1^{13} = \tau_2^{13} = 1, \tau_1^{23} = \tau_2^{23} = 0, \tau_1^{33} = \tau_2^{33} = 1$$

$$P_1 = \begin{pmatrix} 71.5815 & 49.0675 \\ 49.0675 & 78.6195 \end{pmatrix}, P_2 = \begin{pmatrix} 71.5815 & 49.0675 \\ 49.0675 & 78.6195 \end{pmatrix},$$

$$P_3 = \begin{pmatrix} 70.8724 & 14.7524 \\ 14.7524 & 69.1407 \end{pmatrix}$$

which enables us to conclude that the system is asymptotically stable. An example of simulation is given by Figure 2.2 with various initial conditions.

2.5. Robust stability

This section examines the stability of uncertain continuous multiple models. Various conditions which use quadratic Lyapunov functions, as well as a class of nonquadratic functions, are examined. The conditions proposed are in LMI form and are easy to solve using traditional numeric tools.

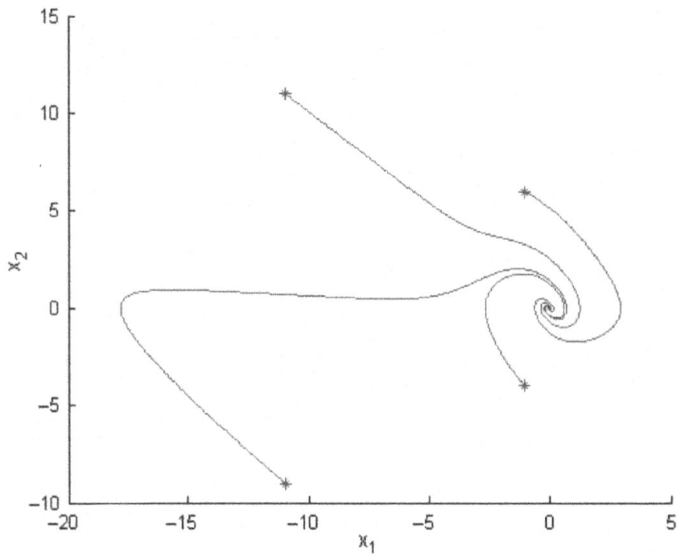

Figure 2.2. *Multiple model [2.24] phase plane with various initial conditions*

We consider the example of time-continuous uncertain multiple models of the form:

$$\dot{x}(t) = \sum_{i=1}^{N} \mu_i(z(t))(A_i + \Delta A_i(t))x(t) \qquad [2.27]$$

where matrices $\Delta A_i(t)$ are unknown and variable over time. Various types of uncertainties can be considered: i) norm-bounded uncertainties, i.e. $\| \Delta A_i(t)) \| \leq \delta_i, \delta_i > 0$, ii) structured uncertainties of the form $\Delta A_i(t) = D_i F_i(t) E_i$, $F_i(t)^T F_i(t) \leq I$ or iii) interval uncertainties, which will be examined in Chapter 5.

The following lemma will be used later [CHA 06a].

LEMMA 2.2.– Let D and E be constant matrices, and F a constant unknown matrix of an appropriate size to satisfy the constraint $F^T F \leq I$, where I is the identity matrix. The following two propositions are equivalent:

– $DFE + E^T F^T D^T < 0$.

– $\varepsilon DD^T + \varepsilon^{-1} E^T E < 0, \ \forall \, \varepsilon > 0$.

2.5.1. *Norm-bounded uncertainties*

Let us first consider norm-bounded uncertainties:

$$\| \Delta A_i(t)) \| \leq \delta_i, \quad \forall \, i \in I_N \qquad [2.28]$$

where δ_i is a positive scalar.

Asymptotic stability of [2.27] can be directly obtained by considering the following conditions:

$$\left(A_i + \Delta A_i \right)^T P + P\left(A_i + \Delta A_i \right) < 0 \qquad \forall i \in I_N$$

which is equivalent to:

$$A_i^T P + PA_i + \Delta A_i^T P + P\Delta A_i < 0 \qquad [2.29]$$

By using Lemma 2.2, with $D = P, F = I_n, \, E = \Delta A_i$, inequalities [2.29] become:

$$A_i^T P + PA_i + \varepsilon_i^{-1} P^2 + \varepsilon_i \delta_i^2 I_n < 0, \quad \varepsilon_i > 0, \forall i \in I_N \qquad [2.30]$$

Effectively, uncertain multiple model [2.27] is asymptotically stable if a positive definite symmetric matrix P and a positive scalar β exist, which are solutions of the following LMI $\forall \, i \in I_N$:

$$\begin{pmatrix} A_i^T P + PA_i + \varepsilon_i \delta_i^2 I_n & P \\ P^T & -\varepsilon_i \end{pmatrix} < 0 \qquad [2.31]$$

This result is obtained by applying the Schur complement onto [2.30].

2.5.2. *Structured parametric uncertainties*

Parametric uncertainties are assumed to be norm-bounded and structured. They satisfy the following property:

$$\Delta A_i(t) = D_i F_i(t) E_i, \quad F_i(t)^T F_i(t) \leq I \qquad [2.32]$$

where D_i, E_i are constant matrices with appropriate dimensions, $F_i(t)$ is an unknown matrix and I is the identity matrix.

Asymptotic stability of [2.27] can be directly obtained by considering:

$$\left(A_i + D_i F_i(t) E_i\right)^T P + P\left(A_i + D_i F_i(t) E_i\right) < 0 \qquad \forall i \in I_N \quad [2.33]$$

By using Lemma 2.2, these matrix inequalities become:

$$A_i^T P + PA_i + \varepsilon_i E_i^T E_i + \varepsilon_i^{-1} PD_i D_i^T P < 0, \; \varepsilon_i > 0$$

Using the Schur complement, the following LMI are obtained:

$$\begin{pmatrix} A_i^T P + PA_i + \varepsilon_i E_i^T E_i & PD_i \\ D_i^T P & -\varepsilon_i I \end{pmatrix} < 0 \qquad [2.34]$$

Effectively, if matrix $P > 0$ and scalars $\varepsilon_i > 0$ exist such that LMI [2.34] are satisfied, then uncertain multiple model [2.27] is asymptotically stable.

These often conservative conditions will be relaxed by extending the conditions of Theorem 2.2 to the robust stability of uncertain multiple models [2.27]. Indeed, the following LMI propose conditions which are sufficient for robust stability.

THEOREM 2.3.– If there exist matrices $P_i > 0$ and scalars τ_k^{ij} and $\varepsilon_{ij} > 0, (i,j) \in I_N^2$ such that:

$$\begin{pmatrix} A_i^T P_j + P_j A_i + \sum_{k=1}^{n} \tau_k^{ij} \left(P_j - P_k\right) + \varepsilon_{ij} E_i^T E_i & P_j D_i \\ * & -\varepsilon_{ij} I \end{pmatrix} < 0 \qquad [2.35]$$

then uncertain multiple model [2.27] is asymptotically stable.

PROOF.– The stability of uncertain multiple model [2.27] is directly obtained from Theorem 2.2:

$$\left(A_i + \Delta A_i\right)^T P_j + P_j \left(A_i + \Delta A_i\right) + \sum_{k=1}^{N} \tau_k^{ij} \left(P_j - P_k\right) < 0 \qquad [2.36]$$

Using Lemma 2.2 and taking into account the uncertainties expression [2.32], inequalities [2.36] are equivalent to:

$$A_i^T P_j + P_j A_i + \sum_{k=1}^{N} \tau_k^{ij} \left(P_j - P_k\right) + \varepsilon_{ij} E_i^T E_i + \varepsilon_{ij}^{-1} P_j D_i D_i^T P_j < 0 \qquad [2.37]$$

By applying the Schur complement, conditions [2.35] are obtained.

We note that for $P_i = P, \forall\, i \in I_N$ we have $P_j - P_k = 0$ and condition [2.37] becomes [2.34]. These stability conditions are consequently less conservative than conditions using a common positive definite matrix. Note also that the N^2 constraints are not simultaneously linear in $P_i > 0$ and τ_k^{ij}. However, if we fix scalars τ_k^{ij}, we obtain a problem which is LMI in $P_i > 0$ and $\varepsilon_{ij} > 0, (i, j) \in I_N^2$.

EXAMPLE 2.2.– Let us consider multiple model [2.27], where $N = 3$, $F_i^T F_i \leq I$, $\Delta A_i = E_i F_i D_i$ and

$$A_1 = \begin{pmatrix} 0 & 1 \\ -0.1 & -1 \end{pmatrix},\ E_1 = \begin{pmatrix} 0 & 1 \\ 0 & -1 \end{pmatrix},\ D_1 = \begin{pmatrix} 0.0667 & 0 \\ 0 & 0.0667 \end{pmatrix}$$

$$A_2 = \begin{pmatrix} 0 & 1.2 \\ -1.9 & -1 \end{pmatrix},\ E_2 = \begin{pmatrix} 0 & 1 \\ -1 & -1 \end{pmatrix},$$

$$D_2 = \begin{pmatrix} 0.0667 & 0 \\ 0 & 0.0667 \end{pmatrix}$$

$$A_3 = \begin{pmatrix} 0 & 2 \\ -0.5 & -1.5 \end{pmatrix},\ E_3 = \begin{pmatrix} 0 & 2 \\ 0 & -1 \end{pmatrix},$$

$$D_3 = \begin{pmatrix} 0.0667 & 0 \\ 0 & 0.0667 \end{pmatrix}$$

Activation functions (Figure 2.3) are:

$$\mu_1(x_1(t)) = \frac{\omega_1(x_1(t))}{\omega_1(x_1(t)) + \omega_2(x_1(t)) + \omega_3(x_1(t))}$$

$$\mu_2(x_1(t)) = \frac{\omega_2(x_1(t))}{\omega_1(x_1(t)) + \omega_2(x_1(t)) + \omega_3(x_1(t))}$$

$$\mu_3(x_1(t)) = \frac{\omega_3(x_1(t))}{\omega_1(x_1(t)) + \omega_2(x_1(t)) + \omega_3(x_1(t))}$$

where:

$$\omega_1(x_1(t)) = \exp\left(-\frac{1}{2}\left(\frac{x_1(t)+c}{d}\right)^2\right),$$

$$\omega_2(x_1(t)) = \exp\left(-\frac{1}{2}\left(\frac{x_1(t)}{d}\right)^2\right),$$

$$\omega_3(x_1(t)) = \exp\left(-\frac{1}{2}\left(\frac{x_1(t)-c}{d}\right)^2\right)$$

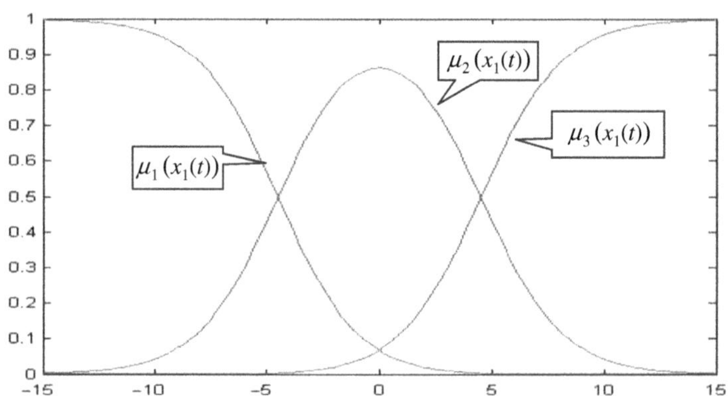

Figure 2.3. *Activation functions, where* $c = 9$, $d = 4$

2.5.3. *Analysis of nominal stability*

Although the three nominal matrices ($\Delta A_i = 0, i \in I_3$) are stable and the simulation shows the stability from the example (see Figure 2.4), the conditions for quadratic stability [2.4] fail to prove the stability of this example. Verification can easily be achieved by solving the problem in

Lemma 2.1, i.e. if there exist matrices $X_i, \forall i \in I_n$, not all null, such that:

$$X_1 > 0, X_2 > 0, X_3 > 0$$

$$X_1 A_1^T + A_1 X_1 + X_2 A_2^T + A_2 X_2 + X_3 A_3^T + A_3 X_3 > 0$$

This problem is solvable and gives:

$$X_1 = \begin{pmatrix} 2041.9 & 639.7 \\ 639.7 & 210.5 \end{pmatrix}, \; X_2 = \begin{pmatrix} 447.4 & -583.7 \\ -583.7 & 768.5 \end{pmatrix},$$

$$X_3 = \begin{pmatrix} 66.5 & 34.5 \\ 34.5 & 27.5 \end{pmatrix}$$

The stability conditions from Theorem 2.2 result in new constraints. Using a method which is iterative over $\tau_k^{ij}, (i,j,k) \in I_n^3$, the problem which is LMI in $P_1 > 0, P_2 > 0$, $P_3 > 0$ leads to the following solution:

$$P_1 = \begin{pmatrix} 149.4433 & 84.1699 \\ 84.1699 & 162.2407 \end{pmatrix}, \; P_2 = \begin{pmatrix} 161.1414 & 99.1047 \\ 99.1047 & 174.9105 \end{pmatrix},$$

$$P_3 = \begin{pmatrix} 151.4240 & 57.5350 \\ 57.5350 & 179.8239 \end{pmatrix}$$

$$\tau_2^{11} = \tau_3^{11} = 0, \; \tau_2^{21} = \tau_3^{21} = 0{,}5, \; \tau_2^{31} = \tau_3^{31} = 0$$

$$\tau_1^{12} = \tau_3^{12} = 0, \; \tau_1^{22} = \tau_3^{22} = 1, \; \tau_1^{32} = \tau_3^{32} = 0$$

$$\tau_1^{13} = \tau_2^{13} = 0.5, \; \tau_1^{23} = \tau_2^{23} = 0, \; \tau_1^{33} = \tau_2^{33} = 0$$

which proves the global asymptotic stability of the nominal multiple model. The simulations from Figure 2.4 illustrate the stable dynamic behavior of the example.

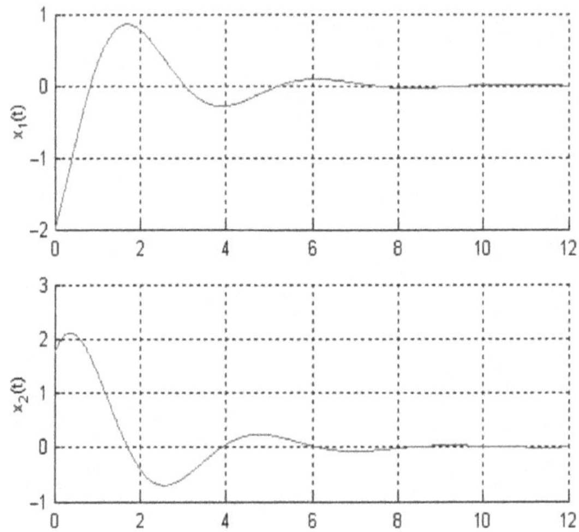

Figure 2.4. *Example simulation of the nominal multiple model with initial conditions* $x(0)^T = (-2, 1, 75)$

2.5.4. *Analysis of robust stability*

Using a method which is iterative over the positive scalar $\tau_k^{ij}, (i, j, k) \in I_N^3$, the stability conditions from Theorem 2.3 result in LMI in $P_1 > 0, P_2 > 0$, $P_3 > 0$ and ε_{ij} $(i, j) \in I_N^2$. If we resolve these LMI, we obtain:

$$P_1 = \begin{pmatrix} 39.57 & 26.39 \\ 26.39 & 47.11 \end{pmatrix}, \ P_2 = \begin{pmatrix} 35.99 & 10.51 \\ 10.51 & 40.90 \end{pmatrix},$$

$$P_3 = \begin{pmatrix} 37.87 & 10.53 \\ 10.53 & 42.87 \end{pmatrix}$$

$$\tau_2^{11} = \tau_3^{11} = 0, \tau_2^{21} = \tau_3^{21} = 1, \tau_2^{31} = \tau_3^{31} = 0$$

$$\tau_1^{12} = \tau_3^{12} = 0.5, \tau_1^{22} = \tau_3^{22} = 0, \tau_1^{32} = \tau_3^{32} = 1$$

$$\tau_1^{13} = \tau_2^{13} = 1, \tau_1^{23} = \tau_2^{23} = 0, \tau_1^{33} = \tau_2^{33} = 1$$

$$\varepsilon_{11}=0.7912,\ \varepsilon_{12}=0.7597,\ \varepsilon_{13}=1.1174$$

$$\varepsilon_{21}=0.2281,\ \varepsilon_{22}=0.2909,\ \varepsilon_{23}=0.3137$$

$$\varepsilon_{31}=0.2810,\ \varepsilon_{32}=0.9569,\ \varepsilon_{33}=0.4369$$

which proves the global asymptotic stability of the uncertain multiple model.

2.6. Conclusion

The stability of continuous-time multiple models, with and without uncertainties, has been studied. Some sufficient conditions for stability have been proposed for quadratic and nonquadratic examples. For nonquadratic cases, stability has led to some interesting results, which – relaxing results from quadratic cases – results, however, in nonconvex constraints. Some conditions for robust stability have also been examined with various types of uncertainties. Stability conditions are LMI in nature and some illustrative examples have been proposed. Other results using polyquadratic functions can be consulted in [CHA 00, CHA 06b, GUE 04, JAD 99, JOH 98, KRU 08, TAN 01]. With regards to polyquadratic functions, it is important to note that it is very difficult to find satisfactory results within the continuous domain.

Chapter 3

Multiple Model State Estimation

3.1. Introduction

This chapter deals with multiple model state estimation. This involves designing observers, also known as multiple observers, which are constructed by interpolation of local Luenberger observers [LUE 64, ORE 83] through the same activation functions as those used for the multiple model [CHA 02b, MUR 97]. Observer synthesis is an essential topic in automation, as synthesis techniques based on full state feedback require a whole state vector to be available, which is rarely verified in practice. These observers relate to linear or nonlinear models which are subjected to unknown inputs [BOU 00, CHA 08e, DAR 94, GUA 91, KOE 02, KUD 80, SHE 03, VAL 99, XIO 03, YAN 88]. Within the multiple model context, we are going to consider for measurable decision variables, multiple models with uncertainties and also unknown inputs [AKH 03, CHA 09, CHA 10, CHA 12a]. Various techniques are used to perform state estimation.

Some relaxation techniques are used throughout this book, for the synthesis of both observers and controllers.

There is literature on various techniques which can easily be used [GUE 04, LIU 03, TAN 98, TUA 01]. In this book, we have chosen relaxations presented in [LIU 03]. Indeed, a matrix constraint of the form $\sum_{i=1}^{N}\sum_{j=1}^{N}\mu_i(z(t))\mu_j(z(t))\Upsilon_{ij} < 0$ is assured if the following satisfactory matrix conditions are verified:

$$\Upsilon_{ij} < 0, \ \forall (i,j) \in I_N^2 \qquad [3.1]$$

This matrix constraint is relaxed by introducing supplementary variables and by separating terms in $i = j$ from terms in $i < j : 1,..,N,$ as shown in the following lemma [LIU 03].

LEMMA 3.1.– If there exist matrices $Q_{ii} > 0$ and Q_{ij} that verify the following conditions $\forall (i,j) \in I_N^2, i < j$:

$$\Upsilon_{ii} + Q_{ii} < 0$$

$$\Upsilon_{ij} + \Upsilon_{ji} + Q_{ij} + Q_{ij}^T < 0$$

$$Q = Q^T = \begin{pmatrix} Q_{11} & Q_{12} & \cdots & Q_{1N} \\ * & Q_{22} & \cdots & Q_{2N} \\ \vdots & \vdots & \ddots & \\ * & * & \cdots & Q_{NN} \end{pmatrix} > 0$$

then constraints [3.1] are also verified.

Relaxation can be intuitively proven by noting that the off-diagonal coefficients of positive definite matrix Q, i.e. matrices Q_{ij} for $i < j$, are not necessarily positive definite. This enables the constraints on the sum $\Upsilon_{ij} + \Upsilon_{ji}$ to be relaxed, being able to be negative semi-definite, compared with the initial constraint [3.1], i.e. $\Upsilon_{ij} < 0, \forall i, j : 1,..,N$. Proof of the resulting relaxation can be found in [LIU 03].

3.2. Synthesis of multiple observers

The control laws proposed are often based on full state feedback. However, in practice, state variables are not always measurable, hence the need for observers. Multiple observers are obtained by interpolation of multiple local Luenberger observers:

$$\begin{cases} \dot{\hat{x}}(t) = \sum_{i=1}^{N} \mu_i(z(t))\left(A_i\hat{x}(t) + B_iu(t) + L_i\left(y(t) - \hat{y}(t)\right)\right) \\ \hat{y}(t) = \sum_{i=1}^{N} \mu_i(z(t))C_i\hat{x}(t) \end{cases} \quad [3.2]$$

where $\hat{x}(t)$ and $\hat{y}(t)$ represent the reconstructed state and output vectors, respectively. It is assumed that the decision variables *z(t)* are uniquely dependent on measurable variables. The activation functions are the same as those of multiple model [2.1] and observe the same properties [2.2]. The estimation error for the state vector is denoted by:

$$e(t) = x(t) - \hat{x}(t) \quad [3.3]$$

The estimation error dynamic is given by the following equation:

$$\dot{e}(t) = \sum_{i=1}^{N}\sum_{j=1}^{N} \mu_i(z(t))\mu_j(z(t))\left(A_i - L_iC_j\right)e(t) \quad [3.4]$$

The aim is to determine the gains $L_i, i \in I_n$ in order to ensure convergence of estimation error toward zero [3.3]. The following theorem proposes a basic result that enables us to calucate the multiple observer [2.1].

THEOREM 3.1.– If there exist matrices $P > 0$ and L_i which verify LMI $\forall i \in I_N$, then:

$$\Sigma_{ii}^T P + P\Sigma_{ii} < 0 \qquad [3.5]$$

$$\left(\Sigma_{ij} + \Sigma_{ji}\right)^T P + P\left(\Sigma_{ij} + \Sigma_{ji}\right) < 0 \qquad [3.6]$$

where $\Sigma_{ij} = A_i - L_i C_j$, then multiple observer [3.1] is globally asymptotically convergent.

PROOF.– By deriving the candidate quadratic function $V\left(e(t)\right) = e(t)^T Pe(t)$, all along the trajectory of system [3.3], we obtain:

$$\dot{V}(e(t)) = e(t)^T \sum_{i=1}^{N}\sum_{j=1}^{N} \mu_i(z(t))\mu_j(z(t))\left(\Sigma_{ij}{}^T P + P\Sigma_{ij}\right)e(t)$$

$$= e(t)^T \sum_{i=1}^{N} \mu_i^2(z)\left(\Sigma_{ii}{}^T P + P\Sigma_{ii}\right) e(t) +$$

$$e(t)^T \sum_{i=1}^{N}\sum_{j>i}^{N} \mu_i(z)\mu_j(z)\left(\left(\Sigma_{ij} + \Sigma_{ji}\right)^T P + P\left(\Sigma_{ij} + \Sigma_{ji}\right)\right)e(t)$$

[3.7]

Conditions [3.5] and [3.6] therefore guarantee the decrease of $V(e(t)) = e(t)^T Pe(t)$, i.e. $\dot{V}(e(t)) < 0$.

3.2.1. *Linearization*

These conditions are in a bilinear matrix inequality (BMI) form in P and L_i. Transformation into linear matrix inequalities (LMI) is easily obtained with the change in variables $PL_i = Y_i$. Effectively, conditions [3.5] and [3.6] become:

$$PA_i - Y_i C_i + \left(PA_i - Y_i C_i\right)^T < 0 \qquad [3.8a]$$

$$P\left(A_i + A_j\right) - Y_j C_i - Y_i C_j + \left(P\left(A_i + A_j\right) - Y_j C_i - Y_i C_j\right)^T < 0 \qquad [3.8b]$$

Solvable LMI are easily obtained using numeric tools.

These conditions are quite conservative as they require stability of all crossed subsystems. The following result enables this conservatism to be alleviated by only imposing stability of dominant submodels (the dominant models are the uncrossed ones) with an additional condition.

THEOREM 3.2.– If there exist matrices $P > 0$, Y_i and Q_{ij} which verify LMI $\forall (i,j) \in I_N^2,\ i < j$:

$$PA_i - Y_i C_i + \left(PA_i - Y_i C_i\right)^T + Q_{ii} < 0 \qquad [3.9a]$$

$$\begin{aligned} &P\left(A_i + A_j\right) - Y_j C_i - Y_i C_j + \left(P\left(A_i + A_j\right) - Y_j C_i - Y_i C_j\right)^T \\ &\qquad + Q_{ij} + Q_{ij}^T < 0 \end{aligned} \qquad [3.9b]$$

$$Q = Q^T = \begin{pmatrix} Q_{11} & Q_{12} & \cdots & Q_{1N} \\ * & Q_{22} & \cdots & Q_{2N} \\ \vdots & \vdots & \ddots & \\ * & * & \cdots & Q_{NN} \end{pmatrix} > 0 \qquad [3.9c]$$

then multiple observer [3.1] is globally asymptotically convergent with $L_i = P^{-1} Y_i$.

Proof results from the derivation of the quadratic function $V\left(e(t)\right) = e(t)^T Pe(t)$, along the trajectory of system [3.3].

Constraint [3.9c] imposes a positive definite symmetric diagonal, i.e. $Q_{ii} > 0$. Conversely, off-diagonal matrices are free, i.e. not necessarily symmetric or positive definite. Consequently, constraints [3.9b] are largely relaxed compared with constraints [3.8b].

Note that these conditions should only be verified for $i < j$ such that $\mu_i(z(t))\mu_j(z(t)) \neq 0$. We recall that for a common output matrix $C_i = C$, estimation error dynamic [3.3] becomes:

$$\dot{e}(t) = \sum_{i=1}^{n} \mu_i(z(t))(A_i - L_i C)e(t) \qquad [3.10]$$

Consequently, results [3.9] are simplified as follows:

$$PA_i - Y_i C + (PA_i - Y_i C)^T < 0 \qquad [3.11]$$

where $L_i = P^{-1}Y_i$.

3.2.2. *Pole placement*

In order to improve multiple observer performances, we can assign poles with the error dynamic [3.10] in a region of the complex plane. Here we examine how to impose specific performance through pole placement in a desired region using LMI techniques [CHI 96]. Firstly, we recall the definition of an LMI region and the formulation of pole placement constraints in an LMI form [CHI 96].

DEFINITION 3.1.– A subset D of the complex plane is said to be an LMI region if there exist matrices $\alpha = (\alpha_{ij}) \in R^{p \cdot p}$ and $\beta = (\beta_{ij}) \in R^{p \cdot p}$ such that $D = \{z \in \mathbb{C} : f_D(z) < 0\}$, where $f_D(z) = (\alpha_{ij} + \beta_{ij}^T z + \beta_{ij}\bar{z})$ and $\mathbb{C}$ is a complex set.

LEMMA 3.2.– A matrix A is said to be D-stable if and only if a positive definite symmetric matrix X exists such that:

$$\alpha \otimes X + \beta \otimes (AX) + \beta^T \otimes (AX)^T < 0$$

Indeed, in order to place the poles from estimation dynamic [3.10] into a particular LMI defined by the two matrices $\alpha = \alpha^T$ and β, which have been previously determined and which are characteristic of the desired performance, it is sufficient to verify:

$$\alpha \otimes P + \beta \otimes (PA_i - Y_i C_i) + \beta^T \otimes (PA_i - Y_i C_i)^T < 0 \text{ with } L_i = P^{-1} Y_i$$

Various LMI regions are proposed in Appendix 1. For example, in order to place the poles in the vertical band defined by Figure 3.1, it is sufficient to verify the following LMI:

$$PA_i - Y_i C_i + (PA_i - Y_i C_i)^T + 2\alpha_1 P < 0$$

$$PA_i - Y_i C_i + (PA_i - Y_i C_i)^T + 2\alpha_2 P > 0$$

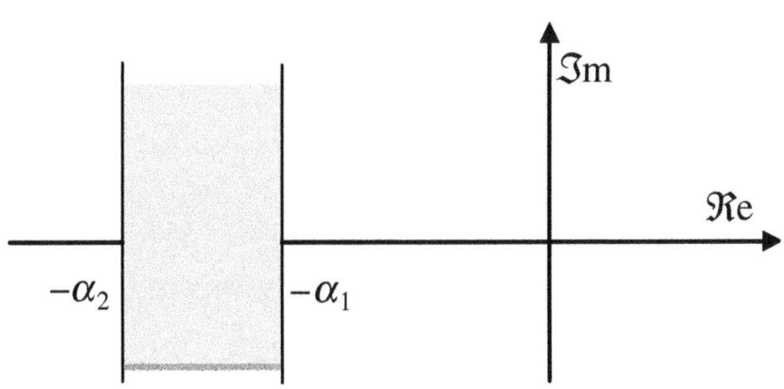

Figure 3.1. *Example of LMI region: vertical band defined by $\alpha_1 > 0$ and $\alpha_2 > 0$*

3.2.3. *Application: asynchronous machine*

Consider an induction motor represented by the following model:

$$\begin{cases} \dot{x}(t) = A(\omega)x(t) + Bu(t) \\ y(t) = Cx(t) \end{cases} \qquad [3.12]$$

where:

$$A(\omega) = \frac{1}{L_r L_s - M^2}\begin{pmatrix} -L_r R_s & \omega M^2 & R_r M & \omega L_r M \\ -\omega M^2 & -L_r R_s & -\omega L_r M & R_r M \\ R_s M & -\omega L_s M & -L_s R_r & -\omega L_r L_s \\ \omega L_s M & R_s M & \omega L_r L_s & -L_s R_r \end{pmatrix}$$

$$B = \frac{1}{L_r L_s - M^2}\begin{pmatrix} L_r & 0 \\ 0 & L_r \\ -M & 0 \\ 0 & M \end{pmatrix}, C = \begin{pmatrix} 1 & 0 & 0 & 0 \\ 0 & 1 & 0 & 0 \end{pmatrix}$$

are state input and output matrices, respectively. The state variables are the stator currents I_{ds} and I_{qs} and the rotor currents I_{dr} and I_{qr}, which are represented by the state vector $x^T = \begin{pmatrix} I_{ds} & I_{qs} & I_{dr} & I_{qr} \end{pmatrix}$. The output variables are represented by the output vector $y^T = \begin{pmatrix} I_{ds} & I_{qs} \end{pmatrix}$, with ω being the rotor speed. The various parameters are given by:

$$R_s = 55;\ R_r = 30;\ L_s = 0.904;\ L_r = 0.904;\ M = 0.842$$

Five operating points around which our system can be linearized are: $\omega_1 = 0$; $\omega_2 = 30$; $\omega_3 = 56.25$; $\omega_4 = 78.3$; $\omega_5 = 93.75$. Consequently, the multiple model will be described by five LTI models:

$$\begin{cases} \dot{x} = \sum_{i=1}^{5} \mu_i(\omega)(A_i x + Bu) \\ y = Cx \end{cases} \qquad [3.13]$$

where A_i represent the state matrices of the multiple model whose numeric values are:

$$A_1 = \begin{pmatrix} -459.2987 & 0.0065 & 233.3444 & 0.0070 \\ -0.0065 & -459.2987 & -0.0070 & 233.3444 \\ 427.7981 & -0.0070 & -250.5265 & -0.0075 \\ 0.0070 & 427.7981 & 0.0075 & -250.5265 \end{pmatrix}$$

$$A_2 = \begin{pmatrix} -459.2987 & 196.4760 & 233.3444 & 210.9434 \\ -196.4760 & -459.2987 & -210.9434 & 233.3444 \\ 427.7981 & -210.9434 & -250.5265 & -226.4760 \\ 210.9434 & 427.7981 & 226.4760 & -250.5265 \end{pmatrix}$$

$$A_3 = \begin{pmatrix} -459.2987 & 368.3925 & 233.3444 & 395.5188 \\ -368.3925 & -459.2987 & -395.5188 & 233.3444 \\ 427.7981 & -395.5188 & -250.5265 & -424.6425 \\ 395.5188 & 427.7981 & 424.6425 & -250.5265 \end{pmatrix}$$

$$A_4 = \begin{pmatrix} -459.2987 & 512.8024 & 233.3444 & 550.5622 \\ -512.8024 & -459.2987 & -550.5622 & 233.3444 \\ 427.7981 & -550.5622 & -250.5265 & -591.1024 \\ 550.5622 & 427.7981 & 591.1024 & -250.5265 \end{pmatrix}$$

$$A_5 = \begin{pmatrix} -459.2987 & 613.9875 & 233.3444 & 659.1980 \\ -613.9875 & -459.2987 & -659.1980 & 233.3444 \\ 427.7981 & -659.1980 & -250.5265 & -707.7375 \\ 659.1980 & 427.7981 & 707.7375 & -250.5265 \end{pmatrix},$$

$$B = \begin{pmatrix} 8.35 & 0 \\ 0 & 8.35 \\ -7.77 & 0 \\ 0 & 7.77 \end{pmatrix}$$

and the activation functions $\mu_i(\omega)$ are defined by:

– for $w \leq w_1$:

$\mu_1(w) = 1,\ \mu_i(w) = 0,\ i = 2,..,5;$

– for $w_1 \leq w \leq w_2$:

$$\mu_1(w) = \frac{w - w_2}{w_1 - w_2},\ \mu_2(w) = 1 - \mu_1(w),\ \mu_i(w) = 0,\ i = 3,4,5;$$

– for $w_2 \leq w \leq w_3$:

$$\mu_2(w) = \frac{w - w_3}{w_2 - w_3},\ \mu_3(w) = 1 - \mu_2(w),\ \mu_i(w) = 0,\ i = 1,4,5;$$

– for $w_3 \leq w \leq w_4$:

$$\mu_3(w) = \frac{w - w_4}{w_3 - w_4},\ \mu_4(w) = 1 - \mu_3(w),\ \mu_1(w) = 0, i = 1,2,5;$$

– for $w_4 \leq w \leq w_5$:

$$\mu_4(w) = \frac{w - w_5}{w_4 - w_5},\ \mu_5(w) = 1 - \mu_4(w),\ \mu_i(w) = 0,\ i = 1,2,3;$$

– for $w \geq w_5$:

$\mu_5(w) = 1,\ \mu_i(w) = 0,\ i = 1,2,3,4.$

The stator currents (I_{ds}, I_{dq}) cannot be measured in practice, whereas the phase currents (I_a, I_b, I_c) can be measured using sensors and can be obtained using the following transformation relation (*Park's inverse transform*):

$$\begin{pmatrix} I_a \\ \hline I_b \\ \hline I_c \end{pmatrix} = T \begin{pmatrix} I_{ds} \\ \hline I_{qs} \end{pmatrix} \quad [3.14]$$

where:

$$T = \begin{pmatrix} 0.8165 & 0 \\ -0.4082 & 0.7071 \\ -0.4082 & -0.7071 \end{pmatrix}$$

Consequently, the new output vector becomes:

$$\bar{y}(t) = T \cdot y(t) = \begin{pmatrix} I_a & I_b & I_c \end{pmatrix}^T \quad [3.15]$$

and the multiple model becomes:

$$\begin{cases} \dot{x}(t) = \sum_{i=1}^{5} \mu_i(\omega)(A_i x(t) + Bu(t)) \\ \bar{y}(t) = TCx(t) = \bar{C}x(t) \end{cases} \quad [3.16]$$

where:

$$\bar{C} = \begin{pmatrix} 0.8165 & 0 & 0 & 0 \\ -0.4082 & 0.7071 & 0 & 0 \\ -0.4082 & -0.7071 & 0 & 0 \end{pmatrix} \quad [3.17]$$

3.2.4. *Synthesis of multiple observers*

We can verify that the pairs $\left(A_i, \bar{C}\right), \forall\, i = 1,\ldots,5,$ are observable:

$$\dot{\hat{x}}(t) = \sum_{i=1}^{5} \mu_i(\omega)(A_i\hat{x}(t) + Bu(t) + L_i(\overline{y}(t) - \overline{C}\hat{x}(t))) \qquad [3.18]$$

We recall that multiple observer [3.18] is exponentially convergent with a rate at least equal to $\alpha > 0$, if matrices Y_i exist such that the following LMI are satisfied:

$$P > 0,\ A_i^T P + PA_i - \overline{C}^T Y_i^T - Y_i\overline{C} + 2\alpha I < 0,\ \ i = 1,..,5 \qquad [3.19]$$

The observer gains are obtained from $L_i = P^{-1}Y_i$.

Solving these LMI with $\alpha = 15$, using Matlab's® LMI *Toolbox*, leads to the following results:

$$P = \begin{pmatrix} 1.7078 & 0 & 0.5663 & 0 \\ 0 & 1.7078 & 0 & 0.5663 \\ 0.5663 & 0 & 0.5647 & 0 \\ 0 & 0.5663 & 0 & 0.5647 \end{pmatrix},$$

$$L_1 = 10^3 \begin{pmatrix} -0.7088 & 0.1769 & 0.0633 \\ -0.0661 & -0.4462 & 0.5109 \\ 1.2904 & -0.1512 & -0.03726 \\ 0.1218 & 0.8225 & -0.7763 \end{pmatrix}$$

$$L_2 = 10^3 \begin{pmatrix} -0.6032 & 0.3330 & 0.1185 \\ -0.0595 & -0.3523 & 0.6049 \\ 0.9317 & -0.8609 & -0.0450 \\ 0.2512 & 0.3440 & -1.2548 \end{pmatrix},$$

$$L_3 = 10^3 \begin{pmatrix} -0.5563 & 0.5113 & 0.0340 \\ -0.3999 & -0.4651 & 0.4920 \\ 1.0262 & -1.1609 & 0.4441 \\ 1.0141 & 0.4235 & -1.1753 \end{pmatrix}$$

$$L_4 = 10^3 \begin{pmatrix} -0,5563 & 0,5113 & 0,0340 \\ -0,3999 & -0,4651 & 0,4920 \\ 1,0262 & -1,1609 & 0,4441 \\ 1,0141 & 0,4235 & -1,1753 \end{pmatrix},$$

$$L_5 = 10^3 \begin{pmatrix} -0,6701 & 0,5688 & -0,2512 \\ -0,4665 & -0,2348 & 0,7223 \\ 1,0404 & -1,5394 & 0,8509 \\ 1,2393 & -0,0315 & -1,6303 \end{pmatrix}$$

Figure 3.2 illustrates the estimation errors $e_i(t) = x_i(t) - \hat{x}_i(t)$, $i = 1,2,3,4$ with initial conditions $x(t) = [0,0,0,0]$, $\hat{x}(t) = [0.5, 0.5, 0.5, 0.5]$.

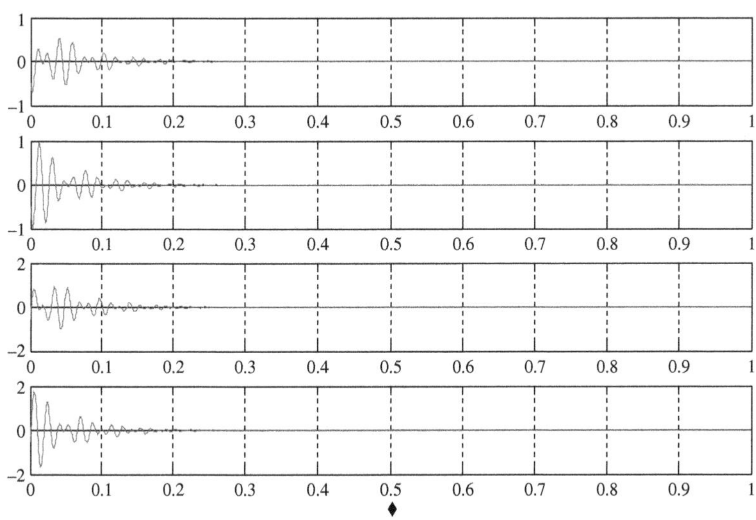

Figure 3.2. *Dynamic of estimation errors*

3.3. Multiple observer for an uncertain multiple model

A model is only a mathematical representation of a system with limited precision. Our aim is to design an observer for uncertain multiple models as follows [AKH 03, AKH 04]:

$$\begin{cases} \dot{x}(t) = \sum_{i=1}^{N} \mu_i\left(z(t)\right)\left(\left(A_i + \Delta A_i(t)\right)x(t) + B_i u(t)\right) \\ y(t) = \sum_{i=1}^{N} \mu_i\left(z(t)\right)C_i x(t) \end{cases} \quad [3.20]$$

where $\Delta A_i(t) \in \mathbb{R}^{n \times n}$ represent the modelization error for state matrix A_i, which is assumed to be norm-bounded:

$$\| \Delta A_i(t) \| < \delta_i \quad [3.21]$$

where $\|.\|$ represents the spectral norm.

The structure of the multiple observer considered is:

$$\begin{cases} \dot{\hat{x}}(t) = \sum_{i=1}^{N} \mu_i\left(z(t)\right)\left(A_i \hat{x}(t) + B_i u(t) + L_i\left(y(t) - \hat{y}(t)\right) + \alpha_i(t)\right) \\ \hat{y}(t) = \sum_{i=1}^{N} \mu_i\left(z(t)\right)C_i \hat{x}(t) \end{cases} \quad [3.22]$$

where $\hat{x}(t) \in \mathbb{R}^n$ is the state estimation and $\hat{y}(t) \in \mathbb{R}^p$ is the output from the observer. $L_i \in \mathbb{R}^{n \times m}$ and $\alpha_i(t) \in \mathbb{R}^n$ are the observer parameters to be determined in order to force the state estimation error to converge toward zero. To that end, we define the state estimation error by:

$$e(t) = x(t) - \hat{x}(t) \quad [3.23]$$

and the output error by:

$$r(t) = y(t) - \hat{y}(t) = \sum_{i=1}^{N} \mu_i(z(t))C_i e(t) \quad [3.24]$$

THEOREM 3.3.– If there exist matrices $P > 0$, W_i and Q_{ij} and positive scalars ε and η which verify the LMI $\forall (i, j) \in I_N^2$, $i < j$:

$$\begin{pmatrix} A_i^T P + PA_i - C_i^T W_i^T - W_i C_i + \delta_i^2 \eta I + Q_{ii} & P \\ * & -\varepsilon I \end{pmatrix} < 0 \qquad [3.25a]$$

$$\left(\begin{array}{c|c} \left(A_i + A_j\right)^T P + P\left(A_i + A_j\right) - C_j^T W_i^T - W_i C_j - & \\ C_i^T W_j^T - W_j C_i + \eta\left(\delta_i^2 + \delta_j^2\right) I + Q_{ij} + Q_{ij}^T & P \\ \hline * & -\varepsilon I \end{array}\right) < 0 \qquad [3.25b]$$

$$Q = Q^T = \begin{pmatrix} Q_{11} & Q_{12} & \cdots & Q_{1N} \\ * & Q_{22} & \cdots & Q_{2N} \\ \vdots & \vdots & \ddots & \\ * & * & \cdots & Q_{NN} \end{pmatrix} > 0 \qquad [3.25c]$$

then observer [3.22] is asymptotically convergent with the following parameters:

$$L_i = P^{-1} W_i \qquad [3.26]$$

and:

$$\begin{cases} \alpha_i(t) = \delta_i^2 \varepsilon \left(1 + \dfrac{\varepsilon}{\eta - \varepsilon}\right) \dfrac{\hat{x}^T(t)\hat{x}(t)}{2r(t)^T r(t)} P^{-1} \displaystyle\sum_{i=1}^{N} \mu_i\left(z(t)\right) C_i^T r(t) \ \ if \ \ r(t) \neq 0 \\ \alpha_i(t) = 0 \ \ if \ \ r(t) = 0 \end{cases}$$

[3.27]

PROOF.– The dynamic of the state estimation error is given by:

$$\dot{e}(t) = \sum_{i=1}^{N} \sum_{j=1}^{N} \mu_i(z) \mu_j(z) \left(\bar{A}_{ij} e(t) + \Delta A_i(t) x - \alpha_i(t)\right) \qquad [3.28]$$

where $\overline{A}_{ij} = A_i - L_i C_j$. The time-derivative of Lyapunov function $V(t) = e^T(t)Pe(t)$ for the whole trajectory of system [3.20] is written as:

$$\dot{V}(t) = \sum_{i=1}^{N}\sum_{j=1}^{N}\mu_i(z(t))\mu_j(z(t))(e^T(t)(\overline{A}_{ij}^T P + P\overline{A}_{ij})e(t)$$

$$+ x^T(t)\Delta A_i^T Pe(t) + e^T(t)P\Delta A_i x(t) - 2\alpha_i^T(t)Pe(t)) \quad [3.29]$$

Using Lemma 2.2, the derivative of Lyapunov function is overvalued as follows:

$$\dot{V}(e(t)) \le \sum_{i=1}^{N}\sum_{j=1}^{N}\mu_i(z(t))\mu_j(z(t))\left(e^T(t)\left(\overline{A}_{ij}^T P + P\overline{A}_{ij} + \varepsilon^{-1}P^2\right)e(t)\right.$$

$$\left. + \delta_i^2 \varepsilon x^T(t)x(t) - 2\alpha_i^T(t)Pe(t)\right) \quad [3.30]$$

which can be rewritten by using [3.28] and by reapplying Lemma 2.2 in the following form:

$$\dot{V}(t) \le \sum_{i=1}^{N}\sum_{j=1}^{N}\mu_i(z(t))\mu_j(z(t))(e^T(t)\left(\overline{A}_{ij}^T P + P\overline{A}_{ij} + \varepsilon^{-1}P^2\right.$$

$$\left. + \delta_i^2 \varepsilon(1+\upsilon)I\right)e(t) + \delta_i^2 \varepsilon(1+\upsilon^{-1})\hat{x}^T(t)\hat{x}(t) - 2\alpha_i^T(t)Pe(t)) \quad [3.31]$$

With the change in variable $\eta = \varepsilon(1+\upsilon)$ in the case where $r(t) \ne 0$ (see equation [3.27]), the following equality can be established:

$$\alpha_i^T(t)Pe(t) = \delta_i^2 \varepsilon(1+\upsilon^{-1})\frac{\hat{x}^T(t)\hat{x}(t)}{2r^T(t)r(t)}r^T(t)\sum_{k=1}^{M}\mu_k(\xi(t))C_k P^{-1}Pe(t)$$

$$= \delta_i^2 \varepsilon(1+\upsilon^{-1})\frac{\hat{x}^T(t)\hat{x}(t)}{2r^T(t)r(t)}r^T(t)r(t) = \frac{1}{2}\delta_i^2 \varepsilon(1+\upsilon^{-1})\hat{x}^T(t)\hat{x}(t) \quad [3.32]$$

Consequently, expression [3.31] can thus be overvalued by:

$$\dot{V}(t) \leq \sum_{i=1}^{N}\sum_{j=1}^{N}\mu_i\left(z(t)\right)\mu_j\left(z(t)\right)e^T(t)\left(\overline{A}_{ij}^T P + P\overline{A}_{ij} + \varepsilon^{-1}P^2\right.$$
$$\left.+\delta_i^2\varepsilon(1+\upsilon)I\right)e(t) \qquad [3.33]$$

By using the Schur complement (see section 1.3.2.2.1) and the variables change $W_i = PL_i$ and $\eta = \varepsilon(1+\upsilon)$, we can show that conditions [3.23]–[3.27] guarantee that $\dot{V}(e(t)) < 0$. In the case where $r(t) = 0$, these conditions directly guarantee that $\dot{V}(e(t)) < 0$.

Note that when multiple model output $y(t)$ is linear, i.e. $C_1 = C_2 = \ldots = C_N = C$, the conditions of Theorem 3.3 are simplified. Indeed, if there exist matrices $P > 0$ and W_i and positive scalars ε and η which verify the LMI $\forall i \in I_N$:

$$\begin{pmatrix} A_i^T P + PA_i - C^T W_i^T - W_i C + \delta_i^2 \eta I & P \\ * & -\varepsilon I \end{pmatrix} < 0 \qquad [3.34]$$

then observer [3.22] is asymptotically convergent with the following parameters:

$$L_i = P^{-1}W_i \qquad [3.35]$$

and:

$$\begin{cases} \alpha_i(t) = \delta_i^2 \varepsilon \left(1 + \dfrac{\varepsilon}{\eta - \varepsilon}\right) \dfrac{\hat{x}^T(t)\hat{x}(t)}{2r(t)^T r(t)} P^{-1}C^T r(t) \;\; si \;\; r(t) \neq 0 \\ \alpha_i(t) = 0 \;\; si \;\; r(t) = 0 \end{cases} \qquad [3.36]$$

Note that in order to simplify the solutions of these LMI, we can fix scalars $\varepsilon = \upsilon = 1$ and consequently $\eta = 2$.

These conditions enable robust observers to be designed taking into account the uncertainties of state matrices. Synthesis of the observer is based on solving LMI conditions knowing *a priori* a boundary δ_i on the uncertainties. However, we note that numeric problems appear during implementation of the observer. Indeed, when the estimation error for output $r(t)$ approaches zero, the measurement $\alpha_i(t)$ approaches infinity. This problem is overcome by slightly modifying parameters $\alpha_i(t)$ by considering a fixed threshold $0 < \kappa << 1$ as follows:

$$\begin{cases} if \ |r_i(t)| > \kappa : \alpha_i(t) = \delta_i^2 \varepsilon \left(1 + \dfrac{\varepsilon}{\eta - \varepsilon}\right) \dfrac{\hat{x}^T(t)\hat{x}(t)}{2r(t)^T r(t)} P^{-1} C^T r(t) \\ if \ |r_i(t)| < \kappa : \alpha_i(t) = 0 \end{cases} \quad [3.37]$$

where $r(t) = (r_1(t), r_2(t), \ldots, r_p(t))^T$. Note that in this case, the state estimation error does not converge toward zero, but remains confined in a neighborhood of zero depending on the choice of $0 < \kappa << 1$.

3.4. Synthesis of unknown input observers

In this section, we examine the synthesis of observers for multiple models subjected to unknown inputs, with and without uncertainties. The objective is the estimation of the state vector and unknown inputs. This type of observer has been widely used in faults diagnosis [CHA 08c, COM 02, HAM 02, ISE 93, SAI 93, TAN 02]. There are various techniques for the synthesis of unknown input observers [CHA 09, CHA 10, UTK 92] such as the high gain technique, by elimination of unknown inputs and using sliding modes. We will only be focusing on the last two techniques. We will show how observer parameters can be determined by solving an LMI set.

In the following, various cases are considered. In the first case, when only the state of the system is affected by unknown inputs, then when unknown inputs also have an effect on system outputs. Finally, the case of uncertain multiple models affected by unknown inputs will be considered.

3.4.1. *Unknown inputs affecting system state*

Let us consider the following multiple model subject to the influence of unknown inputs:

$$\begin{cases} \dot{x}(t) = \sum_{i=1}^{n} \mu_i(z(t))\left(A_i x(t) + B_i u + R_i \bar{u}(t) + D_i\right) \\ y(t) = Cx(t) \end{cases} \quad [3.38]$$

where $x(t) \in \mathbb{R}^n$ is the state vector, $u(t) \in \mathbb{R}^m$ is the input vector, $\bar{u}(t) \in \mathbb{R}^q$ is the unknown inputs vector and $y(t) \in \mathbb{R}^p$ represents the vector of measurable outputs. $A_i \in \mathbb{R}^{n \times n}$ is the state matrix, $B_i \in \mathbb{R}^{n \times m}$ is the known inputs matrix, $R_i \in \mathbb{R}^{n \times q}$ and $D_i \in \mathbb{R}^n$ represent a vector that is dependent on the operating point and $C \in \mathbb{R}^{p \times n}$ is the output matrix. Parameter $z(t)$ represents the decision vector that is dependent on the input and measurable variables.

HYPOTHESIS 3.1.– Matrix $R_i \in \mathbb{R}^{n \times q}$ is of maximum rank, i.e. $rank\left(R_i\right) = q$ and $q < p$.

Let us consider the multiple observer described by:

$$\begin{cases} \dot{\zeta}(t) = \sum_{i=1}^{N} \mu_i\left(z(t)\right)(N_i \zeta(t) + G_{i1} u(t) + G_{i2} + L_i y(t)) \\ \hat{x}(t) = \zeta(t) - Ey(t) \end{cases} \quad [3.39]$$

$N_i \in \mathbb{R}^{n.n}$, $G_{i1} \in \mathbb{R}^{n.m}$, $L_i \in \mathbb{R}^{n.p}$, $G_{i2} \in \mathbb{R}^{n}$ and $E \in \mathbb{R}^{n.p}$ are the observer parameters to be determined. $\hat{x}(t)$ is the estimation of the state of multiple model [3.38].

THEOREM 3.4.– If there exist matrices $X > 0$, S and W_i which verify $\forall i \in I_N$:

$$A_i^T X + XA_i + A_i^T C^T S^T + SCA_i - C^T W_i^T - W_i C < 0 \tag{3.40}$$

$$(X + SC) R_i = 0 \tag{3.41}$$

then multiple observer [3.39] is asymptotically convergent with these parameters given by:

$$\begin{aligned} E &= X^{-1}S \\ G_{i1} &= (I + EC) B_i \\ G_{i2} &= (I + EC) D_i \\ N_i &= (I + EC) A_i - X^{-1} W_i C \\ L_i &= X^{-1} W_i - N_i E \end{aligned} \tag{3.42}$$

PROOF.– Firstly, from [3.41] we have $\dot{\hat{x}}(t) = \dot{\zeta}(t) - EC\dot{x}(t)$. Then, the dynamic for the state estimation error $e(t) = x(t) - \hat{x}(t)$ is expressed as:

$$\dot{e}(t) = J\dot{x}(t) - \dot{\zeta}(t) \tag{3.43}$$

where:

$$J = I + EC \tag{3.44}$$

Taking into account the expressions for multiple model [3.38] and observer [3.39]:

$$\dot{e}(t)=\sum_{i=1}^{N}\mu_i\big(z(t)\big)\Big(N_i e(t)+(JA_i-N_i-K_iC)x(t)+\big(JB_i-G_{i1}\big)u(t)$$
$$+\big(JD_i-G_{i2}\big)+JR\bar{u}(t)\Big)$$

[3.45]

where:

$$K_i=N_iE+L_i \qquad [3.46]$$

Using the variables change:

$$S=XE\ ,\ W_i=XK_i \qquad [3.47]$$

where $X>0$ and constraint [3.41], $JR=0$ is verified. The expressions for gains G_{i1},G_{i2} in [3.42] also verify $JB_i-G_{i1}=0$ and $JD_i-G_{i2}=0$. Consequently, with $N_i=\big(I+EC\big)A_i-K_iC$, equation [3.45] becomes:

$$\dot{e}(t)=\sum_{i=1}^{N}\mu_i\big(z(t)\big)N_ie(t) \qquad [3.48]$$

Indeed, in order to guarantee asymptotic convergence of estimation error dynamic [3.46], it is sufficient that there exists a matrix $X>0$ such that:

$$N_i^TX+XN_i<0,\ \forall i\in I_N \qquad [3.49]$$

With the change in variables [3.47], we obtain conditions [3.42a].

3.4.1.1. *Pole placement*

In order to improve multiple observer performances, we consider the pole placement technique in a complex region [CHI 96]. To illustrate this technique, we consider a particular region of the complex plane $S(\alpha,\beta)$, which is defined by the intersection between a disk, centered at $(0,0)$

and with radius β, and a left half-plane bounded by a straight line with x coordinate $-\alpha < 0$. In order to guarantee the eigenvalues of matrix $\sum_{i=1}^{N} \mu_i(z(t)) N_i$ belonging to region $S(\alpha, \beta)$, it is sufficient to satisfy the following linear conditions:

$$\begin{pmatrix} -\beta X & A_i^T X + A_i^T C^T S^T - C^T W_i^T \\ * & -\beta X \end{pmatrix} < 0 \qquad [3.50a]$$

$$A_i^T X + X A_i + A_i^T C^T S^T + SCA_i - C^T W_i^T - W_i C + 2\alpha X < 0$$

[3.50b]

$$(X + SC) R_i = W_i F \qquad [3.50c]$$

LMI expressions [3.50a] guarantee that the eigenvalues of $\sum_{i=1}^{N} \mu_i(z(t)) N_i$ belong to a disk centered at (0,0) and with radius β. LMI [3.50b] guarantees the belonging of these same eigenvalues to the left half-plane bounded by a straight line x coordinate $-\alpha$ (see Appendix 1).

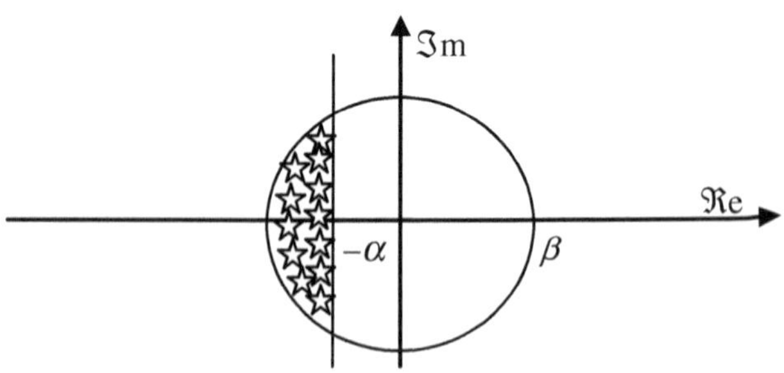

The following section deals with the case where unknown inputs also affect the output equation.

3.4.2. *Unknown inputs affecting system state and output*

Let us consider the following multiple model:

$$\begin{cases} \dot{x}(t) = \sum_{i=1}^{N} \mu_i\left(z(t)\right)\left(A_i x(t) + B_i u(t) + R_i \bar{u}(t) + D_i\right) \\ y(t) = Cx(t) + F\bar{u}(t) \end{cases} \qquad [3.51]$$

where $R_i \in \mathbb{R}^{n \times q}$ and $F \in \mathbb{R}^{p \times q}$ are matrices characterizing the influence of the unknown inputs on the state and output equations, respectively.

By considering the same multiple observer structure [3.39] as previously, observer parameters N_i, L_i, G_{i1}, G_{i2} and E must be determined in order to ensure the asymptotic convergence of estimated state $\hat{x}(t)$ toward multiple model state $x(t)$.

HYPOTHESIS 3.2.–

– $R_i \in \mathbb{R}^{n.q}$ is a full column rank matrix, i.e. $rank\left(R_i\right) = q$ and $q < p$.

– $F \in \mathbb{R}^{p.q}$ is a full column rank matrix, i.e. $rank\left(F\right) = q$.

THEOREM 3.5.– If there exist matrices $X > 0$, S and W_i which verify $\forall i \in I_N$:

$$\begin{aligned} & A_i^T X + XA_i + A_i^T C^T S^T + SCA_i - C^T W_i^T - W_i C < 0 \\ & SF = 0 \\ & \left(X + SC\right) R_i = W_i F \end{aligned} \qquad [3.52]$$

then multiple observer [3.39] is globally asymptotically convergent with the parameters given by [3.42].

PROOF.– The proof is identical to that from the preceding theorem, using the same change in variables $S = XE$, $W_i = XK_i$.

The pole placement in region $S(\alpha,\beta)$, which was previously defined, can be achieved by verifying the following LMI $\forall i \in I_N$:

$$\begin{pmatrix} -\beta X & A_i^T X + A_i^T C^T S^T - C^T W_i^T \\ * & -\beta X \end{pmatrix} < 0$$

$$\begin{aligned} & A_i^T X + XA_i + A_i^T C^T S^T + SCA_i - C^T W_i^T - W_i C + 2\alpha X < 0 \\ & SF = 0 \\ & (X + SC) R_i = W_i F \end{aligned} \qquad [3.53]$$

where X is a positive definite symmetric matrix and S and W_i are matrices with appropriate dimensions.

3.4.3. *Estimation of unknown inputs*

Numerous studies concerning the estimation of unknown inputs of linear dynamic system have been carried out [CHA 09, SAI 93]. One of the most used methods is that based on the good system state estimation. Indeed, if the asymptotic convergence conditions are verified, then, in steady state, the state estimation error approaches zero; we can then replace $x(t)$ by $\hat{x}(t)$ in multiple model equation [3.51]:

$$\begin{cases} \dot{\hat{x}}(t) = \sum_{i=1}^{N} \mu_i(z(t))(A_i \hat{x}(t) + B_i u(t) + R_i \hat{\bar{u}}(t) + D_i) \\ \hat{y}(t) = C\hat{x}(t) + F\hat{\bar{u}}(t) \end{cases} \qquad [3.54]$$

where $\hat{\bar{u}}(t)$ is the only unknown variable. Equation [3.54] can be rewritten as follows:

$$\begin{pmatrix} \dot{\hat{x}}(t) \\ \hat{y}(t) \end{pmatrix} = \begin{pmatrix} \sum_{i=1}^{N} \mu_i\left(z(t)\right)\left(A_i \hat{x}(t) + B_i u(t)\right) \\ C\hat{x}(t) \end{pmatrix} + W\,\hat{\bar{u}}(t) \qquad [3.55a]$$

where:

$$W = \begin{pmatrix} \sum_{i=1}^{N} \mu_i\left(z(t)\right) R_i \\ F \end{pmatrix} \qquad [3.55b]$$

Taking into account Hypothesis 3.2, i.e. W is a full column rank matrix, variable $\hat{\bar{u}}(t)$ is obtained from the pseudo-inverse of matrix W:

$$\hat{\bar{u}}(t) = (W^T W)^{-1} W^T \begin{pmatrix} \dot{\hat{x}}(t) - \sum_{i=1}^{N} \mu_i\left(z(t)\right)\left(A_i \hat{x}(t) + B_i u(t) + D_i\right) \\ y(t) - C\hat{x}(t) \end{pmatrix} \qquad [3.56]$$

As matrix F is assumed to be of full rank, the estimation of the unknown input can also be carried out using the following equation:

$$\hat{\bar{u}}(t) = (F^T F)^{-1} F^T (y(t) - C\hat{x}(t)) \qquad [3.57]$$

EXAMPLE 3.1.– Let us consider the following multiple model:

$$\begin{cases} \dot{x}(t) = \sum_{i=1}^{2} \mu_i(z(t))(A_i x(t) + B_i u(t) + R_i \bar{u}(t)) \\ y(t) = Cx(t) + F\bar{u}(t) + \nu(t) \end{cases} \qquad [3.58]$$

The numeric values of matrices A_i, B_i, R_i, C and F are:

$$A_1 = \begin{pmatrix} -2 & 1 & 1 \\ 1 & -3 & 0 \\ 2 & 1 & -6 \end{pmatrix}, A_2 = \begin{pmatrix} -3 & 2 & 2 \\ 5 & -8 & 0 \\ 0.5 & 0.5 & -4 \end{pmatrix},$$

$$B_1 = \begin{pmatrix} 1 \\ -0.5 \\ -0.5 \end{pmatrix}, B_2 = \begin{pmatrix} -0.5 \\ 1 \\ -0.25 \end{pmatrix}$$

$$F = \begin{pmatrix} 1 \\ 1 \end{pmatrix}, R_1 = \begin{pmatrix} 1 \\ -1 \\ 1 \end{pmatrix}, R_2 = \begin{pmatrix} 1 \\ 0.5 \\ -2 \end{pmatrix}, C = \begin{pmatrix} 1 & 1 & 1 \\ 1 & 0 & 1 \end{pmatrix}$$

Output $y(t)$ is also assumed to be subjected to additive noise $v(t)$, centered, with variance 0.01. The decision variables vector is assumed to be measurable $z(t) = u(t)$.

The multiple observer considered has the following structure:

$$\begin{cases} \dot{\zeta}(t) = \sum_{i=1}^{2} \mu_i \left(z(t)\right)\left(N_i \zeta(t) + G_{i1} u(t) + G_{i2} + L_i y(t)\right) \\ \hat{x}(t) = \zeta(t) - E y(t) \end{cases} \qquad [3.59]$$

Solution of linear constraints [3.52] enables us to determine the observer parameters given by [3.43]. The following numeric values are obtained:

$$N_1 = \begin{pmatrix} -8.25 & -0.69 & -2.62 \\ 0.45 & -13.76 & 0.22 \\ -0.61 & -1.00 & -7.80 \end{pmatrix}, \ N_2 = \begin{pmatrix} -15.82 & -4.74 & 2.31 \\ 1.02 & -14.11 & -0.11 \\ -1.13 & -0.13 & -1.59 \end{pmatrix}$$

$$E = \begin{pmatrix} -2.62 & 2.62 \\ -0.77 & 0.77 \\ -0.80 & 0.80 \end{pmatrix}, G_{11} = \begin{pmatrix} 2.31 \\ -0.11 \\ -0.09 \end{pmatrix}, G_{21} = \begin{pmatrix} -3.12 \\ 0.22 \\ -1.05 \end{pmatrix}$$

$$L_1 = \begin{pmatrix} 33.91 & -30.28 \\ 22.32 & -22.54 \\ 13.13 & -11.32 \end{pmatrix},\ L_2 = \begin{pmatrix} 71.11 & -71.42 \\ 20.58 & -20.47 \\ 11.48 & -13.88 \end{pmatrix},$$

$$X = \begin{pmatrix} 0.1 & 0 & -0.02 \\ 0 & 0.1 & 0 \\ -0.02 & 0 & 0.29 \end{pmatrix}$$

Let us now consider the same model with $y(t) = Cx(t)$, i.e. $F = 0$, solution of LMI [3.42] gives observer parameters [3.52]:

$$X = 10^3 \begin{pmatrix} 1.4190 & 0.2894 & 0.1662 \\ 0.2894 & 0.4869 & 0.1899 \\ 0.1662 & 0.1899 & 0.3893 \end{pmatrix},\quad E = \begin{pmatrix} -0.5 & 0.25 \\ -1 & 1 \\ 0.5 & -1.25 \end{pmatrix}$$

$$G_1 = \begin{pmatrix} 0.375 \\ 0 \\ -0.375 \end{pmatrix},\quad G_2 = \begin{pmatrix} -0.1875 \\ 0 \\ 0.1875 \end{pmatrix}, N_1 = \begin{pmatrix} -1.6968 & 1.9126 & 3.0532 \\ 0.4736 & -8.3044 & 0.4736 \\ -4.5705 & 4.9178 & -9.3205 \end{pmatrix}$$

$$N_2 = \begin{pmatrix} -1.64 & 3.17 & 5.72 \\ -3.71 & -10.90 & -3.71 \\ -3.53 & 9.66 & -10.90 \end{pmatrix},$$

$$L_1 = \begin{pmatrix} -0.37 & 1.43 \\ 0 & 0 \\ 0.37 & -1.43 \end{pmatrix},\ L_2 = \begin{pmatrix} 1.68 & -1.03 \\ 0 & 0 \\ 1.68 & 1.03 \end{pmatrix}$$

Simulation of this example is carried out with initial conditions $x(0) = \begin{pmatrix} 0 & 0 & 0 \end{pmatrix}^T$ and $\hat{x}(0) = \begin{pmatrix} -1 & 0.2 & 0 \end{pmatrix}^T$ and by considering an unknown input $\bar{u}(t)$, as given in Figure 3.4, a unit step control input $u(t)$ and the activation functions from Figure 3.3. Figure 3.5 shows the state variables, the unknown input and their estimations. We are able to observe the proposed observer estimation quality.

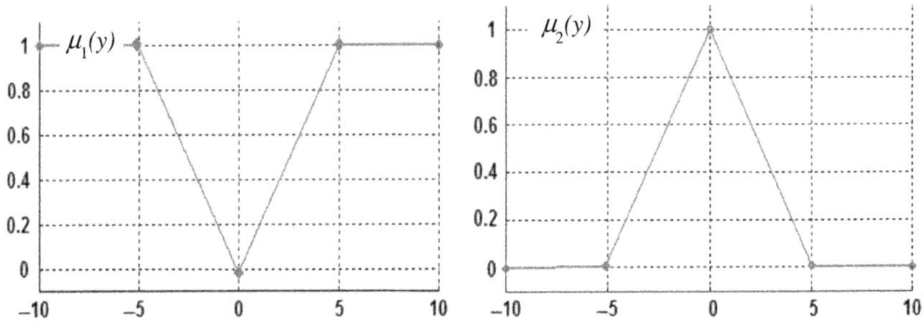

Figure 3.3. *Activation functions $\mu_1(y)$ and $\mu_2(y)$*

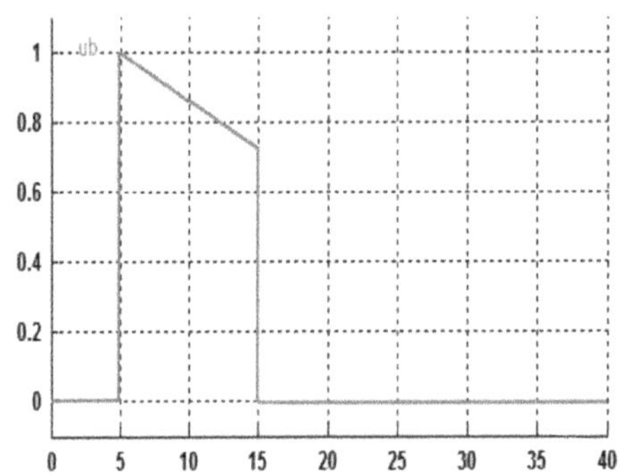

Figure 3.4. *Unknown input $\bar{u}(t)$*

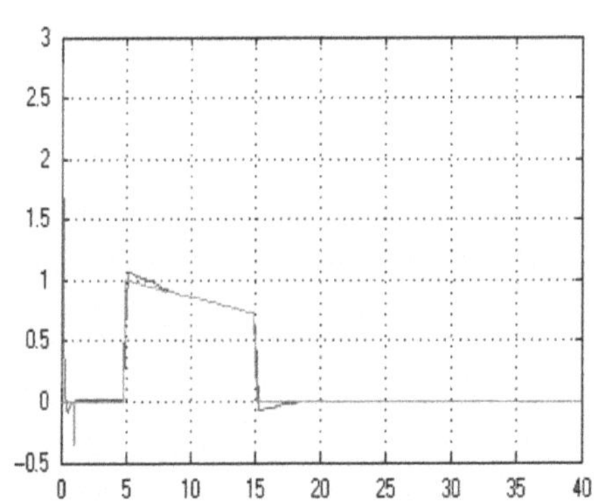

Figure 3.5. *Evolution of the unknown input and its estimation*

3.5. Synthesis of unknown input observers: another approach

The approach presented in this section assumes the unknown inputs vector to be norm-bounded. It is inspired from sliding mode techniques [UTK 92]. The idea is to synthesize the observer parameters, forcing the estimation error dynamic to reach, and remain, around zero bounded by a predefined constant. This technique is used to compensate for both the unknown inputs and model uncertainties.

3.5.1. *Principle*

Let us consider the multiple model described by:

$$\begin{cases} \dot{x}(t) = \sum_{i=1}^{N} \mu_i(z(t))(A_i x(t) + B_i u(t) + R_i \bar{u}(t) + D_i) \\ y(t) = Cx(t) \end{cases} \tag{3.60}$$

HYPOTHESIS 3.3.– The unknown inputs $\bar{u}(t)$ are bounded and nonidentically null, such that $\| \bar{u}(t) \| \leq \rho$, where ρ is a positive scalar and $\| \cdot \|$ represents the Euclidean norm.

The multiple observer structure considered is of the following form:

$$\begin{cases} \dot{\hat{x}}(t) = \sum_{i=1}^{N} \mu_i(z(t))\big(A_i \hat{x}(t) + B_i u(t) + D_i + L_i(y(t) - C\hat{x}(t)) + \alpha_i(t)\big) \\ \hat{y}(t) = C\hat{x}(t) \end{cases} \tag{3.61}$$

Matrices L_i and vectors $\alpha_i(t) \in \mathbb{R}^n$ must be determined to guarantee the asymptotic convergence of $\hat{x}(t)$ toward $x(t)$.

THEOREM 3.6.– If there exist matrices $P>0$ and W_i and a positive scalar β such that $\forall i \in I_N$:

$$A_i^T P + PA_i - C^T W_i^T - W_i C + \beta I < 0 \quad [3.62]$$

then multiple observer [3.61] converges asymptotically and is defined by the following parameters:

$$L_i = P^{-1} W_i \quad [3.63a]$$

$$\alpha_i(t) = \frac{\rho^2 \beta^{-1} \left\| PR_i \right\|^2}{2r^T(t)r(t)} P^{-1} C^T r(t) \;\; if \;\; r(t) \neq 0 \quad [3.63b]$$

$$\alpha_i(t) = 0 \;\; if \;\; r(t) = 0 \quad [3.63c]$$

where

$$r(t) = y(t) - \hat{y}(t).$$

The proof is identical to that of the preceding theorem by considering the state estimation error dynamic described by:

$$\dot{e}(t) = \sum_{i=1}^{N} \mu_i\left(z(t)\right)\left((A_i - G_i C)e(t) + R_i \bar{u}(t) - \alpha_i(t)\right) \quad [3.64]$$

the Lyapunov function $V(e(t)) = e(t)^T Pe(t)$, $P>0$, the change in variables $W_i = PL_i$ and Lemma 2.2.

3.5.1.1. *Extension to different matrices C_i*

In the case of different output matrices C_i, the multiple model is written as follows:

$$\begin{cases} \dot{x}(t) = \sum_{i=1}^{N} \mu_i\left(z(t)\right)\left(A_i x(t) + B_i u(t) + R_i \overline{u}(t) + D_i\right) \\ y(t) = \sum_{i=1}^{N} \mu_i\left(z(t)\right) C_i x(t) \end{cases} \quad [3.65]$$

The structure of the multiple observer used to estimate the state vector is similar to that described in [3.61] by replacing C with C_i.

COROLLARY 3.1.– If there exists a positive scalar β and matrices $P > 0$, Q_{ij} and W_i such that $\forall (i,j) \in I_N^2, i < j$:

$$A_i^T P + PA_i - C_i^T W_i^T - W_i C_i + \beta I + Q_{ii} < 0 \quad [3.66]$$

$$\begin{aligned} &\left(A_i + A_j\right)^T P + P\left(A_i + A_j\right) - C_j^T W_i^T - W_i C_j \\ &\qquad - C_i^T W_j^T - W_j C_i + 2\beta I + Q_{ij} + Q_{ij}^T < 0 \end{aligned} \quad [3.67]$$

$$Q = Q^T = \begin{pmatrix} Q_{11} & Q_{12} & \cdots & Q_{1N} \\ * & Q_{22} & \cdots & Q_{2N} \\ \vdots & \vdots & \ddots & \\ * & * & \cdots & Q_{NN} \end{pmatrix} > 0 \quad [3.68]$$

then multiple observer [3.59] converges asymptotically and is defined by the following parameters:

$$L_i = P^{-1} W_i \quad [3.69a]$$

$$\alpha_i(t) = \frac{\rho^2 \beta^{-1} \left\| PR_i \right\|^2}{2r^T(t) r(t)} P^{-1} \sum_{i=1}^{N} \mu_i(z(t)) C_i^T r(t) \;\; if\; r(t) \neq 0 \quad [3.69b]$$

$$\alpha_i(t) = 0 \;\; if\; r(t) = 0 \quad [3.69c]$$

The same techniques for pole placement in LMI regions can be used.

3.5.2. *Multiple observers subject to unknown inputs and uncertainties*

In this section, we are taking into account the simultaneous presence of unknown inputs and uncertainties:

$$\begin{cases} \dot{x}(t) = \sum_{i=1}^{N} \mu_i\left(z(t)\right)\left(\left(A_i + \Delta A_i(t)\right)x(t) + B_i u(t) + R_i \bar{u}(t) + D_i\right) \\ y(t) = Cx(t) \end{cases} \quad [3.70]$$

We recall that uncertainties on state matrices $\Delta A_i(t)$ and unknown inputs $\bar{u}(t)$ are bounded such that $\|\Delta A_i(t)\| < \delta_i$ and $\|\bar{u}(t)\| \leq \rho$, where δ_i and ρ are positive scalars.

The multiple observer considered has the form:

$$\begin{cases} \dot{\hat{x}}(t) = \sum_{i=1}^{N} \mu_i\left(z(t)\right)\left(A_i \hat{x}(t) + B_i u(t) + D_i + L_i\left(y(t) - C\hat{x}(t)\right) + v_i(t) + \alpha_i(t)\right) \\ \hat{y}(t) = C\hat{x}(t) \end{cases} \quad [3.71]$$

Synthesis of this observer consists of determining gain matrices L_i, variables $v_i(t) \in \mathbb{R}^n$ and $\alpha_i(t) \in \mathbb{R}^n$.

THEOREM 3.7.– If there exist matrices $P > 0$ and W_i and positive scalars β, η and ε such that:

$$\begin{pmatrix} A_i^T P + PA_i - C^T W_i^T - W_i C + \delta_i^2 \eta I + \beta I & P \\ * & -\varepsilon I \end{pmatrix} < 0 \quad [3.72]$$

then multiple observer [3.71] is asymptotically convergent and defined by the following parameters:

$$L_i = P^{-1}W_i \qquad [3.73a]$$

$$\begin{cases} \nu_i(t) = \dfrac{\rho^2 \beta_i^{-1} \|PR_i\|^2}{2r^T(t)r(t)} P^{-1}C^T r(t), \\ \alpha_i(t) = \delta_i^2 \varepsilon \dfrac{\eta}{\eta - \varepsilon} \dfrac{\hat{x}^T(t)\hat{x}(t)}{2r(t)^T r(t)} P^{-1}C^T r(t) \end{cases} \quad \textit{if } r(t) \neq 0 \qquad [3.73b]$$

$$\nu_i(t) = 0 \text{ and } \alpha_i(t) = 0 \textit{ if } r(t) = 0 \qquad [3.73c]$$

The proof is identical to that of the preceding theorem by considering the state estimation error dynamic:

$$\dot{e}(t) = \sum_{i=1}^{N} \mu_i(z(t))\left((A_i - G_iC)e(t) + \Delta A_i(t)x(t) + R_i\bar{u}(t) - \nu_i(t) - \alpha_i(t)\right)$$

and the same change in variables $W_i = PL_i$.

Other relaxed results based on polyquadratic Lyapunov functions and including the case of unmeasurable decision variables can be found in [CHA 12b].

3.6. Conclusion

In this chapter, the state estimation of multiple models is tackled using a variety of techniques. We have considered different multiple observers for multiple models subjected to unknown inputs and uncertainties. These methods have the advantage of being formulated in a convex optimization problem (LMI) and are easily solved using numerical tools. Some simulation examples are given to illustrate these synthesis conditions.

Chapter 4

Stabilization of Multiple Models

4.1. Introduction

The previous chapters have been devoted to analyzing the stability of multiple models and state estimation, based on Lyapunov functions and LMI formulation. In this chapter, we discuss some control law synthesis approaches.

The synthesis of controllers has been actively considered in recent years and numerous studies have been published in this area. These studies, which have been mostly inspired by control techniques found in literature, have dealt with control through static and dynamic full state feedback, with and without uncertainties [CHA 10, CHA 06b, GUE 04, JAD 99, JOH 98, KRU 08, TAN 01, TAN 98]. Multiobjective synthesis techniques have also been tackled.

4.2. Full state feedback control

The continuous closed-loop multiple model via PDC control law [2.40] is given by:

$$\dot{x}(t)=\sum_{i=1}^{N}\sum_{j=1}^{N}\mu_i\big(z(t)\big)\mu_j\big(z(t)\big)\big(A_i-B_iK_j\big)x(t) \qquad [4.1]$$

Synthesis of control laws using full state feedback can be obtained by directly replacing the analysis conditions, state matrices for open-loop multiple models with those relating to closed-loop multiple models, i.e.:

$$\big(A_i-B_iK_j\big)^T P+P\big(A_i-B_iK_j\big)<0 \qquad \forall\ (i,j)\in I_N^2$$

The obtained result has the drawback of being very conservative and not directly formulated as an LMI problem. However, as has already been described, there are multiple methods for reducing it to an LMI problem. Some relaxation techniques have also been introduced, as we will see later.

Also note that these constraints need to be solved $\forall(i,j)\in I_N^2$ verifying $\mu_i(z(t))\mu_j(z(t))\neq 0$. This enables us to reduce the number of constraints when activation functions are bounded support.

THEOREM 4.1.– If there exists a matrix $P>0$ such that $\forall(i,j)\in I_N^2,\ i<j$:

$$G_{ii}^T P+PG_{ii}<0 \qquad [4.2]$$

$$\big(G_{ij}+G_{ji}\big)^T P+P\big(G_{ij}+G_{ji}\big)\leq 0 \qquad [4.3]$$

where $G_{ij}=A_i+B_iK_j$, then multiple model [4.1] derived from its stabilizing control law $u(t)=\sum_{i=1}^{N}\mu_i(z(t))K_ix(t)$ is asymptotically stable.

PROOF.– Let $V(x(t)) = x(t)^T Px(t)$ be a quadratic Lyapunov function. Its time-derivative all along system [4.1] gives:

$$\dot{V}(x(t)) = x(t)^T \sum_{i=1}^{N}\sum_{j=1}^{N} \mu_i(z(t))\mu_j(z(t))\left(G_{ij}{}^T P + PG_{ij}\right)x(t)$$

$$= x(t)^T \sum_{i=1}^{N} \mu_i{}^2(z)\left(G_{ii}{}^T P + PG_{ii}\right)x(t)$$

$$+ x(t)^T \sum_{i=1}^{N}\sum_{j>i}^{N} \mu_i(z)\mu_j(z)\left(\left(G_{ij} + G_{ji}\right)^T P + P\left(G_{ij} + G_{ji}\right)\right)x(t)$$

[4.4]

Conditions [4.2] and [4.3] guarantee that $V(x(t)) = x(t)^T Px(t)$ decreases, i.e. $\dot{V}(x(t)) < 0$.

4.2.1. *Linearization*

Conditions [4.2] and [4.3] are in BMI form in K_i and P, Multiplying both sides by $P^{-1} = X > 0$ and if $N_i = K_i X$, then these BMI become LMI in N_i and X:

$$XA_i^T + A_i X + N_i^T B_i^T + B_i N_i < 0,\ \forall i \in I_N \qquad [4.5]$$

$$XA_i^T + A_i X + XA_j^T + A_j X + N_j^T B_i^T$$

$$+ N_i^T B_j^T + B_i N_j + B_j N_i \leq 0,\ \forall (i,j) \in I_N^2,\ i < j \qquad [4.6]$$

The stabilizing control law is thus defined by:

$$u(t) = \sum_{i=1}^{N} \mu_i(z(t)) K_i x(t),\ K_i = N_i X^{-1} \qquad [4.7]$$

The stability conditions of the above theorem are quite conservative, since they require all crossed subsystems to be stable. The following theorem enables this conservatism to be reduced by only imposing that the dominant subsystems (also known as uncrossed models) are stable with an additional condition.

THEOREM 4.2.– If there exist matrices $X>0$, Y_{ij} and N_i which verify the following LMI $\forall(i,j)\in I_N^2, i<j$:

$$XA_i^T + A_iX + N_i^T B_i^T + B_iN_i + Y_{ii} < 0 \qquad [4.8]$$

$$\begin{aligned} &XA_i^T + A_iX + XA_j^T + A_jX + B_iN_j + B_jN_i \\ &\quad + N_i^T B_j^T + N_j^T B_i^T + Y_{ij} + Y_{ij}^T \leq 0 \end{aligned} \qquad [4.9a]$$

$$Y = Y^T = \begin{pmatrix} Y_{11} & \cdots & Y_{1N} \\ * & \ddots & \vdots \\ * & * & Y_{NN} \end{pmatrix} > 0 \qquad [4.9b]$$

then multiple model [4.1] derived from its control law defined by $u(t) = \sum_{i=1}^{N} \mu_i(z(t))N_iX^{-1}x(t)$ is asymptotically stable.

The proof is obtained by considering the quadratic Lyapunov function defined by $V(x(t)) = x(t)^T Px(t)$, $P>0$ and its derivative for the whole multiple model [4.1]. Conditions [4.9] guarantee the negativity of the derivative of:

$$V(x(t)) = x(t)^T Px(t) \qquad [4.10]$$

i.e. $\dot{V}(x(t)) < 0$.

These results enable the conservatism, introduced by the interactions, to be further reduced. The introduction of matrices Y_{ij} that are not necessarily symmetric or positive definite enables condition [4.7] to be relaxed. Other intermediate results are not presented here, but can be found under references [CHA 00, GUE 04, JAD 99, JOH 98, KRU 08, LIU 03, TAN 01]. An illustrative example is proposed at the end of this section.

4.2.2. *Specific case*

In the case where the input matrices verify the condition $B_i = \alpha_i B,\ \alpha_i > 0,\ i:1...N$, use of the following control law:

$$u(t) = \frac{\sum_{i=1}^{N} \mu_i\left(z(t)\right)\alpha_i K_i}{\sum_{i=1}^{N} \mu_i\left(z(t)\right)\alpha_i} x(t) \qquad [4.11]$$

leads to stability conditions that are less conservative than those corresponding to the commonly used control $u(t) = \sum_{i=1}^{N} \mu_i(z(t))K_i x(t)$. The closed-loop multiple model, *via* [4.11], is thus expressed by:

$$\dot{x}(t) = \sum_{i=1}^{N} \mu_i\left(z(t)\right)\left(A_i + B_i K_i\right)x(t) \qquad [4.12]$$

Note that the multiple model obtained is expressed without crossed terms $(i \neq j)$. The stability conditions previously presented are reduced to the existence of a matrix $X > 0$ such that:

$$XA_i^T + A_i X + N_i^T B_i^T + B_i N_i < 0,\ \forall i \in I_N \qquad [4.13]$$

Note that if pairs (A_i, B_i) are controllable, then gains K_i exist that enable eigenvalues of $(A_i - B_i K_i)$ to be placed at the same point. In this case, there is a greater likelihood that a common matrix $P > 0$ exists to all these pairs.

Other results exist that use other types of Lyapunov functions or other techniques [BEN 78]. For example, in [ELK 99], by using the Borne–Gentina criterion [BOR 93a, GEN 72] over the comparative systems, stabilization conditions are proposed for the multiple models of the form:

$$\dot{x}(t) = \sum_{i=1}^{N} \mu_i(z(t)) A_i x(t)) + Bu(t) \qquad [4.14]$$

with state matrices in companion form:

$$A_i = \begin{pmatrix} 0 & 1 & 0 & \dots & 0 \\ \vdots & \ddots & \ddots & \ddots & \vdots \\ \vdots & \ddots & \ddots & \ddots & 0 \\ 0 & \dots & \dots & 0 & 1 \\ -a_{i,0} & \dots & \dots & -a_{i,n-2} & -a_{i,n-1} \end{pmatrix}, B = \begin{pmatrix} 0 \\ \vdots \\ \vdots \\ 0 \\ 1 \end{pmatrix} \qquad [4.15]$$

$a_{i,j}$ are the coefficients of the characteristic polynomial $P_{A_i}(\lambda)$ of matrix A_i such that:

$$P_{A_i}(\lambda) = \lambda^n + \sum\nolimits_{j=1}^{n-1} a_{i,j} \lambda^j \qquad [4.16]$$

with the change-of-basis of the form:

$$T = \begin{pmatrix} 1 & 1 & \dots & 1 & 0 \\ \alpha_1 & \alpha_2 & \dots & \alpha_{n-1} & 0 \\ \vdots & \vdots & \dots & \ddots & \vdots \\ \alpha_1^{n-2} & \alpha_2^{n-2} & \dots & \alpha_{n-1}^{n-2} & 0 \\ \alpha_1^{n-1} & \alpha_2^{n-1} & \dots & \alpha_{n-1}^{n-1} & 1 \end{pmatrix} \qquad [4.17]$$

the following arrow form matrix is obtained [BEN 78]:

$$F_I = T^{-1} A_i T = \begin{pmatrix} \alpha_1 & & & \beta_1 \\ & \ddots & & \vdots \\ & & \alpha_{n-1} & \beta_{n-1} \\ \gamma_{i,1} & \cdots & \gamma_{i,n-1} & \gamma_{i,n} \end{pmatrix} \quad [4.18]$$

where:

$$\beta_j = \prod_{\substack{k=1 \\ k \neq j}}^{n-1} \left(\alpha_j - \alpha_k\right)^{-1}, \ j:1,2,\ldots,n-1$$

$$\gamma_{i,j} = -P_{A_i}\left(\alpha_j\right), \ j:1,2,\ldots,n-1 \quad [4.19]$$

$$\gamma_{i,n} = -a_{i,n-1} - \sum_{i=1}^{n-1} \alpha_i, \ i:1,2,\ldots,n-1$$

where α_i, $i:1,2,\ldots,n-1$ are distinct and arbitrary negative constant parameters. The following theorem summarizes these conditions [BOR 08].

THEOREM 4.3.– The multiple model described in [4.14] is globally asymptotically stable via control law $u(t) = Kx(t)$, if the following condition is satisfied:

$$\gamma'_{i,n} - \sum_{k=1}^{n-1} \frac{|\beta_k| |\gamma'_{i,k}|}{\alpha_k} < 0 \quad [4.20]$$

where:

$$\beta_k = \prod_{\substack{k=1 \\ k \neq j}}^{n-1} \left(\alpha_k - \alpha_j\right)^{-1}, \ k:1,2,\ldots,n-1 \quad [4.21a]$$

$$\gamma'_{i,k} = -P'_{A_i - BK}\left(\alpha_k\right), \ k:1,2,\ldots,n-1$$

$$\gamma'_{i,n} = trace\left(A_i - BK\right) - \sum_{j=1}^{n-1} \alpha_j \quad [4.21b]$$

EXAMPLE 4.1.– To illustrate the degree of conservatism between the above results (Theorems 4.1 and 4.2), let us consider the following example of two subsystems parameterized by par $a, b \in R$:

$$A_1 = \begin{pmatrix} 2 & -10 \\ 1 & 0 \end{pmatrix}, \quad B_1 = \begin{pmatrix} 1 \\ 0 \end{pmatrix} \quad [4.22a]$$

$$A_2 = \begin{pmatrix} a & -10 \\ 1 & 1 \end{pmatrix}, \quad B_2 = \begin{pmatrix} b \\ 0 \end{pmatrix} \quad [4.22b]$$

Figures 4.1 and 4.2 show the parameter domain for which stabilization is assured. It is quite evident that Theorem 4.2 introduces more relaxation.

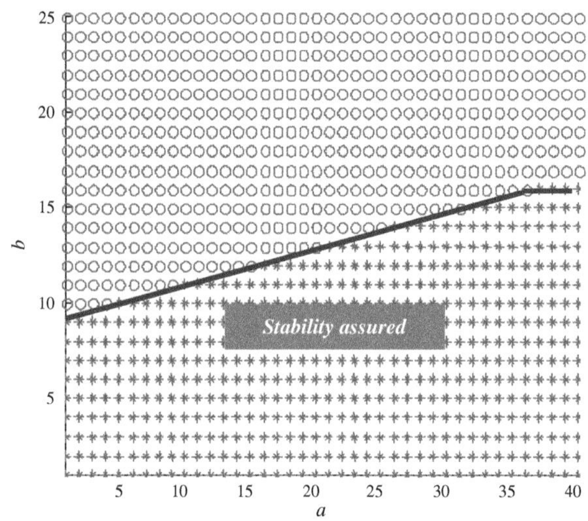

Figure 4.1. *Stability domain (*) – Theorem 4.1*

4.2.3. *α-stability: decay rate*

The above results only guarantee the stability of the multiple model without any performance criterion. The

α-stability criterion that guarantees a certain degree of decay is often used.

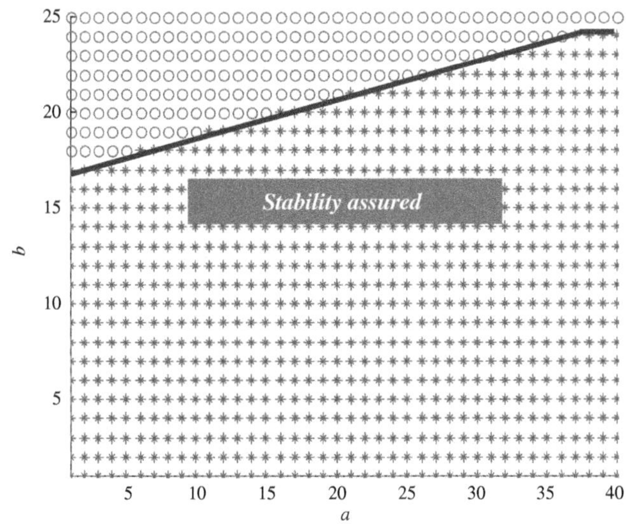

Figure 4.2. *Stability domain (*) – Theorem 4.2 (less conservative than Theorem 4.1)*

THEOREM 4.4.– If there exist matrices $P>0$, Q_{ij}, K_i and a scalar $\alpha \geq 0$ which verify $\forall i<j \in I_N$:

$$G_{ii}^T P + PG_{ii} + \alpha P + Q_{ii} < 0 \qquad [4.23a]$$

$$\left(G_{ij}+G_{ji}\right)^T P + P\left(G_{ij}+G_{ji}\right) + 2\alpha P + Q_{ij} + Q_{ij}^T \leq 0 \qquad [4.23b]$$

$$Q = Q^T = \begin{pmatrix} Q_{11} & Q_{12} & \cdots & Q_{1N} \\ * & Q_{22} & \cdots & Q_{2N} \\ \vdots & \vdots & \ddots & \\ * & * & \cdots & Q_{NN} \end{pmatrix} > 0 \qquad [4.23c]$$

where $G_{ij} = A_i + B_i K_j$, then multiple model [4.1] derived from its stabilizing control law defined by $u(t) = \sum_{i=1}^{N} \mu_i(z(t)) K_i x(t)$ is exponentially stable.

PROOF.– In order to ensure exponential decay of coefficient $\alpha \geq 0$, it is sufficient to examine the stability of $\dot{x}(t) = \sum_{i=1}^{N} \mu_i(z(t))((A_i + \alpha I)x(t)) + B_i u(t))$, where $u(t) = \sum_{i=1}^{N} \mu_i(z(t))$ $K_i x(t)$.

This result regarding the decay rate imposes supplementary constraints that involve dominant (G_{ii}) and crossed (G_{ij}) terms. These additive constraints on the crossed terms lead, quite clearly, to more restrictive design conditions.

The following result enables us to reach this objective by only modifying the dominant terms, i.e. constraints [4.23a]. Constraints [4.23b] are relaxed by taking into account the maximum number (r) of local models that are simultaneously activated. This is equal to the maximum number of local models when the activation functions have an infinite support, i.e. $r = N$ (Figure 4.3 shows an example of Gaussian functions, where $r = N = 3$).

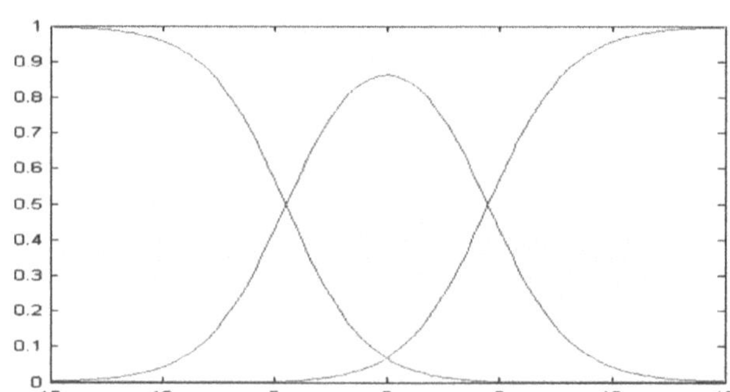

Figure 4.3. *Example of activation functions with infinite support*

To improve the above result, by reducing its conservatism, the following proposition is used [CHA 02b].

PROPOSITION 4.1.– Taking into account the properties of activation functions, the following condition is verified $\forall r \in \{2,..,r\}$:

$$\sum_{i=1}^{N} \mu_i\left(z(t)\right)^2 \geq \frac{1}{r}$$

where r is the maximum number of local models activated at each instant.

PROOF.– The properties of the activation functions enable us to write:

$$1 = \sum_{i=1}^{N} \mu_i\left(z(t)\right)^2 = \sum_{i=1}^{N} \mu_i\left(z(t)\right)^2 + \sum_{i \neq j=1}^{N} \mu_i\left(z(t)\right)\mu_j\left(z(t)\right)$$

Using the following inequality [TAN 98]:

$$\sum_{i=1}^{N} \mu_i\left(z(t)\right)^2 \geq \frac{1}{r-1} \sum_{i \neq j=1}^{N} \mu_i\left(z(t)\right)\mu_j\left(z(t)\right)$$

this proposition is demonstrated.

THEOREM 4.5.– Let r be the maximum number of local models simultaneously activated. If there exist matrices $P > 0$, Q_{ij} and K_i and a scalar $\alpha \geq 0$ which verify $\forall i < j \in I_N$:

$$G_{ii}^T P + PG_{ii} + Q_{ii} + 2\alpha r P < 0 \qquad [4.24a]$$

$$\left(G_{ij} + G_{ji}\right)^T P + P\left(G_{ij} + G_{ji}\right) + Q_{ij} + Q_{ij}^T \leq 0 \qquad [4.24b]$$

$$Q = Q^T = \begin{pmatrix} Q_{11} & Q_{12} & \cdots & Q_{1N} \\ * & Q_{22} & \cdots & Q_{2N} \\ \vdots & \vdots & \ddots & \\ * & * & \cdots & Q_{NN} \end{pmatrix} > 0 \qquad [4.24c]$$

where $G_{ij} = A_i + B_i K_j$, then multiple model [4.1] derived from its stabilizing control law defined by $u(t) = \sum_{i=1}^{N} \mu_i(z(t)) K_i x(t)$ is asymptotically stable.

PROOF.– If constraints [4.24a] and [4.24b] are multiplied by $\mu_i(z(t))^2$ and $\mu_i(z(t))\mu_j(z(t))$, respectively, then by adding them together the following is obtained:

$$\sum_{i=1}^{N} \mu_i\left(z(t)\right)^2 \left(G_{ii}^T P + P G_{ii} + Q_{ii} + 2\alpha P\right)$$

$$+ \sum_{i<j:1}^{N} \mu_i\left(z(t)\right) \mu_j\left(z(t)\right) \Big(\left(G_{ij} + G_{ji}\right)^T P$$

$$+ P\left(G_{ij} + G_{ji}\right) + Q_{ij} + Q_{ij}^T \Big) < 0$$

which allows us to write:

$$G(z)^T P + P G(z) + \bar{\mu}^T Q \bar{\mu} + 2 r \alpha P \sum_{i=1}^{N} \mu_i\left(z(t)\right)^2 < 0 \qquad [4.25]$$

where $G(z) = \sum_{i=1}^{N} \sum_{j=1}^{N} \mu_i\left(z(t)\right) \mu_j\left(z(t)\right) G_{ij}$, $\bar{\mu}^T = \left(\mu_1(.), \mu_2(.), \ldots, \mu_n(.)\right)$.

Knowing that $\sum_{i=1}^{N} \mu_i^2(z) \geq \frac{1}{r}$ (Proposition 4.1), we can write:

$$G(z)^T P + P G(z) + 2\alpha P < 0 \qquad [4.26]$$

which corresponds to $\dot{V}\left(x(t)\right)+2\alpha V\left(x(t)\right)<0$, where $V\left(x(t)\right)=x(t)^T Px(t),\ P>0$ is the corresponding Lyapunov function.

Note that the maximum decay rate can be obtained by considering the GEVP problem in $X>0$ and α:

Maximize α such that:

$$\begin{aligned} & XA_i^T + A_iX + N_i^T B_i^T + B_iN_i + Y_{ii} + 2r\alpha X < 0 \\ & XA_i^T + A_iX + XA_j^T + A_jX + N_i^T B_i^T + B_iN_i \\ & \qquad + N_i^T B_j^T + B_jN_i + Y_{ij} + Y_{ij}^T \leq 0 \\ & \begin{pmatrix} Y_{11} & \cdots & Y_{1N} \\ * & \ddots & \vdots \\ * & * & Y_{NN} \end{pmatrix} > 0 \end{aligned} \qquad [4.27]$$

where $Y_{ij} = XQ_{ij}X$ and $K_i = N_iX^{-1}, \forall\ (i,j)\in I_N^2,\ i<j.$

Note also that the same result can be obtained by considering other relaxation techniques. The following example illustrates the synthesis of the control law $u(t)=\sum_{i=1}^{N}\mu_i(z(t))K_ix(t)$ with a certain degree of stability.

EXAMPLE 4.2.– Let us consider multiple model [4.22] from Example 4.1 with $a=10, b=5$ and activation functions:

$$\mu_1\left(x_1(t)\right)=\frac{1-\tanh\left(x_1(t)\right)}{2},\ \mu_2\left(x_1(t)\right)=\frac{1+\tanh\left(x_1(t)\right)}{2} \qquad [4.28]$$

Closed-loop multiple model [4.22] is written as:

$$\dot{x}(t)=\sum_{i=1}^{2}\sum_{i=1}^{2}\mu_i\left(x(t)\right)\mu_j\left(x(t)\right)\left(A_i - B_iK_j\right)x(t) \qquad [4.29]$$

Synthesis of this control law with degree of stability $\alpha = 15$ leads to:

$$P = \begin{pmatrix} 0.0078 & 0.377 \\ 0.377 & 24.3090 \end{pmatrix}, Q = \begin{pmatrix} 0.4 & 19.5 \\ 19.5 & 1092.4 \end{pmatrix}$$

$$K_1 = (170.4 \quad 8307.5), K_2 = (49 \quad 2325.4) \qquad [4.30]$$

Figure 4.4 shows the range evolution of eigenvalues of matrix $G(x(t)) = \sum_{i=1}^{2} \sum_{i=1}^{2} \mu_i(x(t))\mu_j(x(t))(A_i - B_i K_j)$. It verifies that the real parts of the eigenvalues of matrix $G(x(t))$ are all to the left of -15.

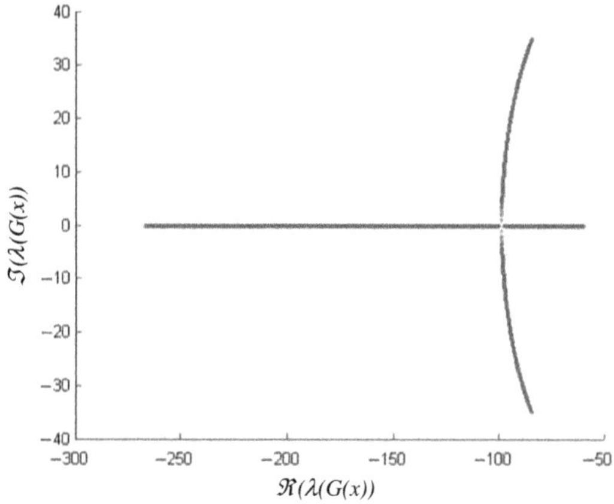

Figure 4.4. *The range evolution of eigenvalues of matrix G(x)*

The full state feedback control assumes the complete availability of the state variables of the system, which is not always possible. We are therefore forced to consider problems where the control is uniquely dependent on the available variables (the measured output $y(.)$ of the system and its inputs). In this sense, two techniques are considered: i) the observer-based control, and ii) static output feedback control.

4.3. Observer-based controller

The observer considered is the one described in Chapter 3 by:

$$\begin{cases} \dot{\hat{x}}(t) = \sum_{i=1}^{N} \mu_i\left(z(t)\right)(A_i\hat{x}(t) + B_i u(t) + L_i\left(y(t) - \hat{y}(t)\right)) \\ \hat{y}(t) = \sum_{i=1}^{N} \mu_i\left(z(t)\right) C_i \hat{x}(t) \end{cases} \quad [4.31]$$

The considered controller is as follows:

$$u(t) = \sum_{i=1}^{N} \mu_i\left(z(t)\right) K_i \hat{x}(t) \quad [4.32]$$

The estimation error dynamic $e(t) = x(t) - \hat{x}(t)$ is written as:

$$\dot{e}(t) = \sum_{i=1}^{N}\sum_{j=1}^{N} \mu_i\left(z(t)\right)\mu_j\left(z(t)\right)\left(A_i - L_i C_j\right)e(t) \quad [4.33]$$

Taking into account control law [4.32] and the estimation error, observer [4.31] becomes:

$$\dot{\hat{x}}(t) = \sum_{i=1}^{N}\sum_{i=1}^{N} \mu_i\left(z(t)\right)\mu_j\left(z(t)\right)\left(\left(A_i + B_i K_j\right)\hat{x}(t) + L_i C_j e(t)\right) \quad [4.34]$$

The improved model is expressed by:

$$\dot{\bar{x}}(t) = \sum_{i=1}^{N}\sum_{i=1}^{N} \mu_i\left(z(t)\right)\mu_j\left(z(t)\right)\bar{G}_{ij}\bar{x}(t) \quad [4.35]$$

where:

$$\bar{G}_{ij} = \begin{pmatrix} A_i + B_i K_j & L_i C_j \\ 0 & A_i - L_i C_j \end{pmatrix} \quad [4.36]$$

$$\overline{x}(t)=\left(\hat{x}(t)^T \quad \tilde{x}(t)^T\right)^T \qquad [4.37]$$

The stability conditions for augmented multiple model [4.35] can easily be obtained by replacing matrices $G_{ij} = A_i + B_i K_j$ with $\overline{G}_{ij}$ in above results [4.2] and [4.3]. The conditions obtained are nonlinear based on variables P, Q, L_i and K_i. They are LMI-transformable by considering diagonal Lyapunov matrices. The separation property is also guaranteed in this case, as the following result shows.

THEOREM 4.6.– If there exist matrices $X > 0$, $Q > 0$, Y_{ij}, Q_{ij}, M_i and N_i which verify the following LMI $\forall (i,j) \in I_N^2, i < j$:

$$XA_i^T + A_i X + N_i^T B_i^T + B_i N_i + Y_{ii} < 0 \qquad [4.38]$$

$$X\left(A_i^T + A_j^T\right) + \left(A_i + A_j\right)X + B_i N_j + B_j N_i$$
$$+N_i^T B_j^T + N_j^T B_i^T + Y_{ij} + Y_{ij}^T \leq 0 \qquad [4.39]$$

$$\begin{pmatrix} Y_{11} & \cdots & Y_{1N} \\ * & \ddots & \vdots \\ * & * & Y_{NN} \end{pmatrix} > 0 \qquad [4.40]$$

$$QA_i + A_i^T Q^T - Y_i C_i - C_i^T Y_i^T + Q_{ii} < 0 \qquad [4.41]$$

$$Q\left(A_i + A_j\right) + \left(A_i + A_j\right)^T Q^T - Y_j C_i - Y_i C_j$$
$$-C_i^T Y_j^T - C_j^T Y_i^T + Q_{ij} + Q_{ij}^T < 0 \qquad [4.42]$$

$$\begin{pmatrix} Q_{11} & \cdots & Q_{1N} \\ * & \ddots & \vdots \\ * & * & Q_{NN} \end{pmatrix} > 0 \qquad [4.43]$$

then multiple model [4.35] derived from its control law defined by $u(t)=\sum_{i=1}^{N}\mu_i(z(t))K_i\hat{x}(t)$, $K_i = N_i X^{-1}$ and its observer [4.31] defined by $L_i = P^{-1}Y_i$ is asymptotically stable.

Proof of this result is obtained by considering a Lyapunov matrix of the form:

$$P=\begin{pmatrix} P_1 & 0 \\ 0 & P_2 \end{pmatrix} \qquad [4.44]$$

with $P_1 = X^{-1}$, $P_2 = Q$ and traditional changes in variable.

These conditions enable the observer and controller to be designed separately, while still guaranteeing the stability of the closed-loop system (separation principle). Note also that in order to obtain an observer with a convergence rate to define, we can use conditions [4.27].

4.3.1. *Unmeasurable decision variables*

In some cases, some or all of the decision variable vectors cannot be measured. Consequently, the activation functions of the control law, as well as of the observer, differ from those of the multiple model. In this case, the decision variables that are dependent on estimated state variables $\hat{x}(t)$ are denoted by $\hat{z}(t)$. Indeed, multiple observer [4.31] and the control law are rewritten:

$$\begin{cases} \dot{\hat{x}}(t)=\sum_{i=1}^{N}\mu_i\left(\hat{z}(t)\right)\left(A_i\hat{x}(t)+B_iu(t)+L_i\left(y(t)-\hat{y}(t)\right)\right) \\ \hat{y}(t)=\sum_{i=1}^{N}\mu_i\left(\hat{z}(t)\right)C_i\hat{x}(t) \end{cases} \qquad [4.45]$$

where:

$$u(t)=\sum_{i=1}^{N}\mu_i\left(\hat{z}(t)\right)K_i\hat{x}(t) \qquad [4.46]$$

The augmented system is then described by:

$$\dot{\bar{x}}(t)=\sum_{i=1}^{N}\sum_{j=1}^{N}\sum_{h=1}^{N}\mu_i\left(z(t)\right)\mu_j\left(\hat{z}(t)\right)\mu_h\left(\hat{z}(t)\right)\tilde{G}_{ijh}\bar{x}(t) \qquad [4.47]$$

where $\bar{x}(t)=\begin{pmatrix}\hat{x}(t)^T & \tilde{x}(t)^T\end{pmatrix}^T$ and:

$$\tilde{G}_{ijh}=\begin{pmatrix} A_i+B_iK_h & -B_iK_h \\ A_i-A_j+\left(B_i-B_j\right)K_h+L_j\left(C_h-C_i\right) & A_j-L_jC_h-\left(B_i-B_j\right)K_h \end{pmatrix} \qquad [4.48]$$

From this, we are able to deduce the synthesis conditions for the observer and the controller by applying the same techniques as those previously presented:

$$\tilde{G}_{ijj}^T P+P\tilde{G}_{ijj}<0 \qquad [4.49]$$

$$\left(\tilde{G}_{ijh}+\tilde{G}_{ihj}\right)^T P+P\left(\tilde{G}_{ijh}+\tilde{G}_{ihj}\right)<0 \qquad [4.50]$$

$\forall i,j,h\in I_N$, $j<h$ with $P>0$. The conditions obtained are nonlinear and are not LMI-transformable using traditional change in variable techniques. A sequential resolution method can be used by considering:

$$P=\begin{pmatrix} P_1 & 0 \\ 0 & P_2 \end{pmatrix} \qquad [4.51]$$

Substituting [4.51] into [4.49] and [4.50] respectively, we obtain:

$$\begin{pmatrix} G_{ij}^T P_1 + P_1 G_{ij} & P_1 B_i K_j + S_{ijj}^T P_2 \\ * & \Xi_{ijj} \end{pmatrix} < 0 \qquad [4.52]$$

$$\begin{pmatrix} \left(G_{ih} + G_{ij}\right)^T P_1 + P_1\left(G_{ih} + G_{ij}\right) & P_1 B_i\left(K_j + K_h\right) + \left(S_{ijh} + S_{ihj}\right)^T P_2 \\ * & \Xi_{ijh} + \Xi_{ihj} \end{pmatrix} < 0$$

$$\forall i, j, h \in I_N,\ j < h \qquad [4.53]$$

where:

$$G_{ij} = A_i + B_i K_j$$

$$\Xi_{ijh} = \left(A_j - L_j C_h + \left(B_i - B_j\right) K_h\right)^T P_2 + P_2\left(A_j - L_j C_h + \left(B_i - B_j\right) K_h\right)$$

$$S_{ijh} = \Delta A_{ij} + \Delta B_{ij} K_h + L_j \Delta C_{hi}$$

Unfortunately, these matrix inequalities are always BMI, dependent on variables P_1, P_2, L_i and K_i, which are difficult to solve simultaneously. One method used to solve this problem consists of sequentially solving the conditions affecting the control law gains (i.e. elements (1,1) of the block matrices [4.52] and [4.53]) and then later on the observer parameters. The process for calculating L_i and K_i is as follows [CHA 02b]:

1) Calculate P_{10} and K_{i0} from:

$$G_{ij}^T P_1 + P_1 G_{ij} < 0 \qquad [4.54]$$

$$\left(G_{ih} + G_{ij}\right)^T P_1 + P_1\left(G_{ih} + G_{ij}\right) < 0 \qquad [4.55]$$

$\forall i,j,h \in I_N,\ j<h$. If a solution exists, then go to step 2, otherwise stop.

2) Matrices P_{10} and K_{i0} are integrated into constraints [4.52] and [4.53], becoming LMI in P_2 and $Y_i = P_2 L_i$. If a solution exists, then go to step 3, otherwise stop.

3) Having obtained P_2 and Y_i, deduce $L_i = Y_i P_2^{-1}$.

Other linearization techniques that are more complex to implement may also be used [HAS 99].

EXAMPLE 4.3.– Let us consider the multiple model from Example 4.1 with a = 49, b = 10 and:

$$\mu_1(x_1(t)) = 1 - \mu_2(x_1(t)) = \begin{cases} 1 - \dfrac{|x_1(t)|}{3}, \forall x_1(t) \in [-3,3] \\ 0, \text{otherwise} \end{cases}$$

From condition [4.38], we determine:

$K_1 = (4.92 \quad -0.53)$, $K_2 = (5.18 \quad -0.09)$

$$P_1 = \begin{pmatrix} 0.028 & 0.043 \\ 0.043 & 0.147 \end{pmatrix}, Q_1 = \begin{pmatrix} 0.032 & 0.064 \\ 0.064 & 0.171 \end{pmatrix}$$

Condition [4.39] also enables us to calculate:

$L_1 = (3.0398 \quad -9.0539)^T$, $L_2 = (50.0398 \quad -9.0539)^T$

$$P_2 = \begin{pmatrix} 29.9427 & 1.5509 \\ 1.5509 & 29.9427 \end{pmatrix}, Q_2 = \begin{pmatrix} 8.4517 & 0.00 \\ 0.00 & 8.4275 \end{pmatrix}$$

We can easily verify that conditions [4.38] and [4.39] guarantee the exponential global stability of the augmented system. A simulation example is given in Figure 4.5 with initial conditions $x(0) = (1.7, -0.7)^T$ and $\hat{x}(0) = (1.5, -0.5)^T$.

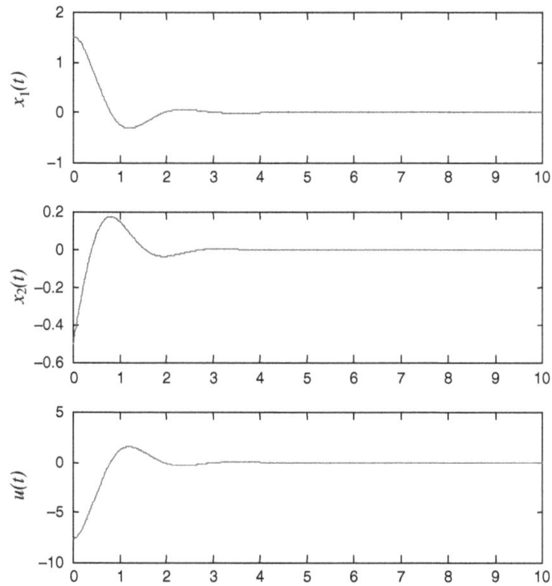

Figure 4.5. *Closed-loop system*

4.4. Static output feedback control

This section deals with the problem of stabilization by static output feedback, which is reputed to be an *NP*-hard problem. We examine an approach that relies on an LMI transformation of the BMI problem obtained. The problem of pole placement is also tackled. Let us consider the following multiple model:

$$\begin{cases} \dot{x}(t) = \sum_{i=1}^{N} \mu_i\left(z(t)\right)\left(A_i x(t) + B_i u(t)\right) \\ y(t) = Cx(t) \end{cases} \qquad [4.56]$$

Considering the control law of the form:

$$u(t) = \sum_{i=1}^{N} \mu_i\left(z(t)\right) F_i y(t) \qquad [4.57]$$

the closed-loop multiple model is expressed as:

$$\dot{x}(t) = \sum_{i=1}^{N}\sum_{j=1}^{N}\mu_i(z(t))\mu_j(z(t))\bar{A}_{ij}\,x(t) \qquad [4.58]$$

where:

$$\bar{A}_{ij} = A_i + B_i F_j C \qquad [4.59]$$

HYPOTHESIS 4.1.– C is a full column rank matrix.

Synthesis of this control law may be directly obtained by substituting matrix $G_{ij} = A_i - B_i K_j$ by matrix $\bar{A}_{ij}$ in the state space feedback synthesis conditions, i.e. for example:

$$P > 0,\, Q > 0 \qquad [4.60a]$$

$$\bar{A}_{ii}^T P + P\bar{A}_{ii} + Q_{ii} < 0 \qquad [4.60b]$$

$$\left(\bar{A}_{ij} + \bar{A}_{ji}\right)^T P + P\left(\bar{A}_{ij} + \bar{A}_{ji}\right) + Q_{ij} + Q_{ij}^T \leq 0 \qquad [4.60c]$$

$$\begin{pmatrix} Q_{11} & \cdots & Q_{1N} \\ * & \ddots & \vdots \\ * & * & Q_{NN} \end{pmatrix} > 0 \qquad [4.60d]$$

These conditions are bilinear in P and F_i $\forall i \in I_N$ and impossible to linearize using traditional change in variable techniques. A linear LMI formulation under linear algebraic constraints is proposed from the following result [CHA 02b].

THEOREM 4.7.– If there exist symmetric matrices $X > 0$, Y_{ii} and matrices Y_{ij}, N_i and M which verify the following conditions $\forall (i,j) \in I_N^2,\, i < j$:

$$A_iX + XA_i^T + B_iN_iC + C^T N_i^T B_i^T + Y_{ii} < 0 \qquad [4.61a]$$

$$X\left(A_i^T + A_j^T\right) + \left(A_i + A_j\right)X + B_iN_jC + B_jN_iC$$

$$+ C^T N_i^T B_j^T + C^T N_j^T B_i^T + Y_{ij} + Y_{ij}^T \le 0 \qquad [4.61b]$$

$$\begin{pmatrix} Y_{11} & \cdots & Y_{1N} \\ * & \ddots & \vdots \\ * & * & Y_{NN} \end{pmatrix} > 0 \qquad [4.61c]$$

$$CX = MC \qquad [4.61d]$$

then the augmented closed-loop multiple model [4.58] via control law $u(t) = \sum_{i=1}^{N} \mu_i(z(t)) F_i y(t),\ F_i = N_i M^{-1}$ is asymptotically stable.

Note that in the case of linear output feedback $F_i = F$, these conditions are reduced to the existence of matrices $X > 0$, N and M such that:

$$A_iX + XA_i^T + B_iNC + C^T N^T B_i^T < 0 \qquad [4.62]$$

$$CX = MC \qquad [4.63]$$

The parameters of the control law are defined by $F = NM^{-1}$. Note that as C is assumed to be a full column rank matrix, it can be deduced that $M = CXC^T\left(CC^T\right)^{-1}$ where $F_i = N_iM^{-1}$. Note also that the obtained constraints are linear and easy to implement. From the numeric point of view, these conditions are easy to solve by using existing numeric tools, such as LMITOOL.

In order to guarantee a certain rate of decay, the proposed result of Theorem 4.5 can easily be extended to the result of Theorem 4.7. Indeed, it is sufficient to replace LMI [4.61a] by:

$$A_i X + X A_i^T + B_i N_i C + C^T N_i^T B_i^T + Y_{ii} + 2r\alpha X < 0 \qquad [4.64]$$

The other constraints remain unchanged.

4.4.1. *Pole placement*

The performances improvement of the static output feedback control is inspired from the state space feedback. Indeed, in order to place the poles of a closed-loop system [4.58] in a particular LMI region defined by two previously determined matrices $\alpha = \alpha^T$ and β, which characterize the desired performance, it is sufficient that:

$$M_D\left(\bar{A}_{ij}, X\right) = \alpha \otimes X + \beta \otimes \left(\bar{A}_{ij} X\right) + \beta^T \otimes \left(\bar{A}_{ij} X\right)^T < 0 \qquad [4.65]$$

where $\bar{A}_{ij} = A_i + B_i F_j C$, which is equivalent to the existence of a symmetric matrix $X > 0$ and matrices N_i and M which verify:

$$\alpha \otimes X + \beta \otimes \left(A_i X + B_i N_j C\right) + \beta^T \otimes \left(A_i X + B_i N_j C\right)^T < 0 \qquad [4.66]$$

$$CX = MC \qquad [4.67]$$

where:

$$F_i = N_i M^{-1} \qquad [4.68]$$

The asymptotic stability of closed-loop multiple model [4.58] is guaranteed with the desired performances.

In the case of a linear static output feedback control $(F_i = F)$, it is sufficient to replace N_j by N in [4.66]. Note also that in the specific case of positive colinearity of input matrices, the following control law is more interesting than the static output feedback control [4.57] [CHA 02b, CHA 07]:

$$u(t) = \frac{\sum_{i=1}^{N} \mu_i(z(t))\alpha_i F_i}{\sum_{i=1}^{N} \mu_i(z(t))\alpha_i} y(t) \qquad [4.69]$$

Multiple model [4.56] where $B_i = \alpha_i B, \alpha_i > 0$ derived from control law [4.69] is written as:

$$\begin{aligned}\dot{x}(t) &= \sum_{i=1}^{N} \mu_i(z(t)) \left(A_i + \alpha_i B \frac{\sum_{i=1}^{N} \mu_i(z(t))\alpha_i F_i}{\sum_{i=1}^{N} \mu_i(z(t))\alpha_i} C \right) x(t) \\ &= \left(\sum_{i=1}^{N} \mu_i(z(t)) A_i + \sum_{i=1}^{N} \mu_i(z(t))\alpha_i B \frac{\sum_{i=1}^{N} \mu_i(z(t))\alpha_i F_i}{\sum_{i=1}^{N} \mu_i(z(t))\alpha_i} C \right) x(t) \\ &= \left(\sum_{i=1}^{N} \mu_i(z(t)) A_i + B \sum_{i=1}^{N} \mu_i(z(t))\alpha_i F_i C \right) x(t)\end{aligned}$$

which gives:

$$\dot{x}(t) = \sum_{i=1}^{N} \mu_i(z(t)) \bar{A}_{ii} x(t) \qquad [4.70]$$

where $\bar{A}_{ii} = A_i + B_i F_i C$. The synthesis conditions for this control law become:

$$A_i X + X A_i^T + B_i N_i C + C^T N_i^T B_i^T < 0, \forall i \in I_N \qquad [4.71]$$

$$CX = MC \qquad [4.72]$$

Interest in this control law stems from the fact that only $(n+1)$ linear constraints are required, instead of (n^2+1) for nonlinear control law [4.57].

EXAMPLE 4.4.– Let us consider the following multiple model:

$$\dot{x}(t) = \sum_{i=1}^{2} \mu_i(y(t))(A_i x(t) + B_i u(t))$$

$$y(t) = Cx(t) \qquad [4.73]$$

where:

$$A_1 = \begin{pmatrix} 2 & -10 \\ 1 & 0 \end{pmatrix}, \quad B_1 = \begin{pmatrix} 1 \\ 0 \end{pmatrix} \qquad [4.74]$$

$$A_2 = \begin{pmatrix} 1 & -10 \\ 1 & 0 \end{pmatrix}, \quad B_2 = \begin{pmatrix} 10 \\ 0 \end{pmatrix} \qquad [4.75]$$

$$C = \begin{pmatrix} 1 & 0 \end{pmatrix} \qquad [4.76]$$

$$\mu_1(y(t)) = \frac{1 - \tanh(y(t))}{2}, \quad \mu_2(y(t)) = \frac{1 + \tanh(y(t))}{2} \qquad [4.77]$$

To illustrate the pole placement problem, let us consider a widely used complex plane region, in which a disk is centered at $(-q, 0)$, with radius $0 < r < q$ (Figure 4.6).

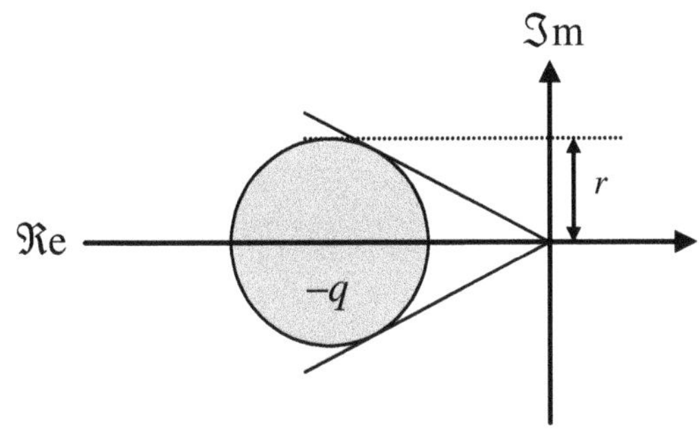

Figure 4.6. *Disk region D_d*

The corresponding LMI region is (see Appendix 1):

$$f_D(z)=\begin{pmatrix} -r & z+q \\ \bar{z}+q & -r \end{pmatrix} \quad [4.78]$$

The advantage of this region is its capacity to fix the decay rate, $-q+r$, and the damping coefficient, $\xi_{\min}=\sqrt{1-r^2/q^2}$ $(r<q)$, of the closed-loop system.

Since $B_2=\alpha B_1, \alpha=10$, the recommended control law to be used is the one described in [4.69]:

$$u(t)=\frac{\mu_1(y(t))F_1+\mu_2(y(t))\alpha F_2}{\mu_1(y(t))+\mu_2(y(t))\alpha}y(t) \quad [4.79]$$

Consequently, the LMI region that corresponds to [4.78] is described (see Appendix 1) by:

$$\begin{pmatrix} -rX & qX+A_iX+B_iN_iC \\ * & -rX \end{pmatrix}<0,\ i:1,2 \quad [4.80a]$$

$$CX=MC \quad [4.80b]$$

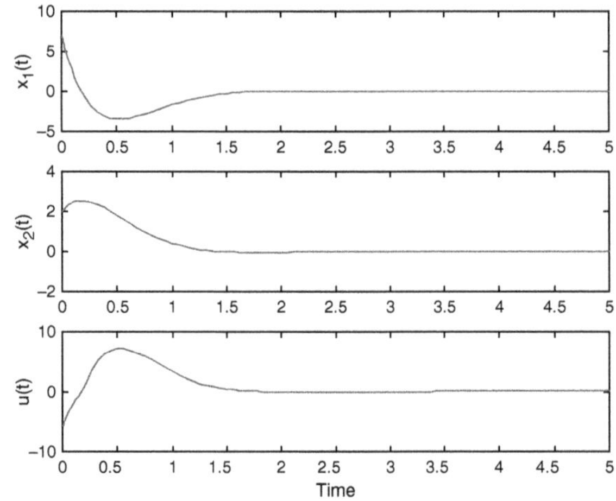

Figure 4.7. *Dynamic of state variables of closed-loop multiple model [4.73] via control law [4.79] with initial conditions x(0)= (7, 2)*

Solving these linear constraints with $q = 10$ and $r = 9$ leads to the following gains:

$$F_1 = -6.9638, F_2 = -0.6094 \qquad [4.81]$$

Simulation results of closed-loop multiple model [4.73] via control law [4.79] is given by Figure 4.7.

4.5. Conclusion

This chapter has proposed sufficient design conditions for various control laws by using quadratic Lyapunov functions. First, LMI synthesis conditions for state space feedback control laws have been determined. Then we have examined an observer-based control. The separation principle has been established in this case. The problem of unmeasurable decision variables is also discussed. Finally, static output feedback controls (OPDC control law) have been presented under an LMI formulation. Some illustrative examples have been proposed.

Chapter 5

Robust Stabilization of Multiple Models

5.1. Introduction

This chapter is essentially devoted to the synthesis of robust control laws for uncertain multiple models. We consider two types of uncertainties: structural norm-bounded and interval uncertainties. Sufficient conditions for the stabilization of uncertain multiple models based on the existence of quadratic Lyapunov functions are proposed. The aim is to present synthesis methods for state–space feedback control, static output feedback and observer-based control laws [CHA 06a, CHA 07]. These synthesis conditions are expressed in an LMI form.

The uncertain multiple model considered is as follows:

$$\begin{aligned} \dot{x}(t) &= \sum_{i=1}^{N} \mu_i(z(t))\left(\hat{A}_i x(t) + \hat{B}_i u(t)\right) \\ y(t) &= Cx(t) \end{aligned} \qquad [5.1]$$

where $\sum_{i=1}^{N} \mu_i(z(t)) = 1, \mu_i(z(t)) \geq 0$, N is the number of local models, $x(t) \in \mathbb{R}^n$ is the state vector, $u(t) \in \mathbb{R}^m$ is the control vector, $y(t) \in \mathbb{R}^l$ is the output vector, and $\hat{A}_i \in \mathbb{R}^{n.n}$, $\hat{B}_i \in \mathbb{R}^{n.m}$, $C_i \in \mathbb{R}^{l.n}$ are the ith state, input and output matrices, respectively. Vector $z(t)$ is the decision variables vector which is dependent on measurable variables.

Two approaches are examined:

– One approach handles the case of structured and norm-bounded uncertainties of the form: $\hat{A}_i = A_i + \Delta A_i$, $\hat{B}_i = B_i + \Delta B_i$, where ΔA_i and ΔB_i are time-variant matrices representing parametric uncertainties verifying:

$$\begin{aligned} &\Delta A_i = D_i F_i(t) E_{Ai}, \quad \Delta B_i = D_i F_i(t) E_{Bi} \\ &F_i(t)^T F_i(t) \leq I \end{aligned} \qquad [5.2]$$

where D_i, E_{Ai}, E_{Bi} are constant matrices of appropriate dimensions, $F_i(t)$ is an unknown matrix obeying constraint [5.2] and I is the identity matrix.

– Another approach considers interval uncertainties such that:

$$\hat{A}_i \in [\underline{A}_i, \overline{A}_i], \ \hat{B}_i \in [\underline{B}_i, \overline{B}_i] \qquad [5.3]$$

i.e. $\hat{A}_i = \{a_{uv}\} \in R^{n.n}$ is such that $[\underline{A}_i, \overline{A}_i] = \{a_{uv} : \underline{a}_{uv} \leq a_{uv} \leq \overline{a}_{uv}\}$ and $\hat{B}_i = \{b_{uv}\} \in R^{n.m}$ is such that $[\underline{B}_i, \overline{B}_i] = \{b_{uv} : \underline{b}_{uv} \leq b_{uv} \leq \overline{b}_{uv}\}$.

These two approaches use Lemma 2.2 (see Chapter 2).

5.2. State feedback control

5.2.1. *Norm-bounded uncertainties*

The considered uncertain multiple model is described by:

$$\begin{aligned}\dot{x}(t) &= \sum_{i=1}^{N} \mu_i(z(t))\big((A_i + \Delta A_i)x(t) + (B_i + \Delta B_i)u(t)\big) \\ y(t) &= \sum_{i=1}^{N} \mu_i(z(t))C_i x(t)\end{aligned} \qquad [5.4]$$

Let us consider the state feedback control law used in earlier chapters:

$$u(t) = \sum_{i=1}^{N} \mu_i(z(t))K_i x(t) \qquad [5.5]$$

where $K_i \in \mathbb{R}^{m \cdot n}$ represents the parameters of the control law to be determined.

The closed-loop multiple model is written as:

$$\dot{x}(t) = \sum_{i=1}^{N}\sum_{j=1}^{N} \mu_i(z(t))\mu_j(z(t))\left(A_i + \Delta A_i + \left(B_i + \Delta B_i\right)K_j\right)x(t) \qquad [5.6]$$

Using the relaxed conditions described above, this model is asymptotically stable if there exist matrices $P > 0$ and K_i such that the following conditions are verified $\forall (i,j) \in I_N^2$, $i < j$:

$$\begin{aligned}P > 0,\ & P\left(A_i + \Delta A_i + \left(B_i + \Delta B_i\right)K_j\right)^T \\ & + \left(A_i + \Delta A_i + \left(B_i + \Delta B_i\right)K_j\right)P < 0\end{aligned} \qquad [5.7]$$

These nonlinear conditions are transformed into linear constraints by applying the techniques used in Lemma 2.2.

THEOREM 5.1.– If there exist matrices N_i and $P>0$ and positive scalars ε_{ij}, κ_{ij} such that the following conditions are verified $\forall(i,j)\in I_N^2$:

$$\Sigma_{ij} = \begin{pmatrix} T_{ij} & * & * \\ E_{Ai}P & -\varepsilon_{ij}I & * \\ E_{Bi}N_j & 0 & -\kappa_{ij}I \end{pmatrix} < 0 \qquad [5.8a]$$

where:

$$T_{ij} = PA_i^T + A_iP + N_j^T B_i^T + B_iN_j + \varepsilon_{ij}D_iD_i^T + \kappa_{ij}D_iD_i^T \qquad [5.8b]$$

then the multiple model [5.6] derived from its control law $u(t)=\sum_{i=1}^{N}\mu_i\left(z(t)\right)K_ix(t)$ with $K_i = N_iP^{-1}, i:1,..,N$ is asymptotically stable.

PROOF.– Using Lemma 2.2, conditions [5.7] are equivalent to:

$$T_{ij} + \varepsilon_{ij}^{-1}PE_{Ai}^TE_{Ai}P + \kappa_{ij}^{-1}N_jE_{Bi}^TE_{Bi}N_j < 0 \qquad [5.9]$$

with T_{ij} defined in [5.8] and where ε_{ij}, κ_{ij} are positive scalars. By applying the Schur complement to [5.9], we obtain:

$$\begin{pmatrix} T_{ij} + \kappa_{ij}^{-1}N_jE_{Bi}^TE_{Bi}N_j & * \\ E_{Ai}P & -\varepsilon_{ij}I \end{pmatrix} < 0 \qquad [5.10]$$

By reapplying the Schur complement to [5.10], [5.8] is obtained.

Relaxation of conditions [5.8] can easily be carried out by using Lemma 3.1 (see Chapter 3) if there exist matrices $Q_{ii} > 0$ and Q_{ij} that verify the following conditions $\forall\, i < j \in I_N$:

$$P > 0,\ \Sigma_{ii} + Q_{ii} < 0 \qquad [5.11a]$$

$$\Sigma_{ij} + \Sigma_{ji} + Q_{ij} + Q_{ij}^T \leq 0 \qquad [5.11b]$$

$$Q = Q^T = \begin{pmatrix} Q_{11} & \cdots & Q_{1N} \\ * & \ddots & \vdots \\ * & * & Q_{NN} \end{pmatrix} > 0 \qquad [5.11c]$$

5.2.2. *Interval uncertainties*

This approach proposes robust stability and stabilization conditions depending on the minimum and maximum bounds of each element of the state and input matrices. The matrices described in [5.3] can be written as [CHA 07]:

$$\underline{A}_i = \begin{pmatrix} \underline{a}_{11i} & \cdots & \underline{a}_{1ni} \\ \vdots & \ddots & \vdots \\ \underline{a}_{1ni} & \cdots & \underline{a}_{1ni} \end{pmatrix}, \overline{A}_i = \begin{pmatrix} \overline{a}_{11i} & \cdots & \overline{a}_{1ni} \\ \vdots & \ddots & \vdots \\ \overline{a}_{1ni} & \cdots & \overline{a}_{1ni} \end{pmatrix} \qquad [5.12a]$$

$$\underline{B}_i = \begin{pmatrix} \underline{b}_{11i} & \cdots & \underline{b}_{1mi} \\ \vdots & \ddots & \vdots \\ \underline{b}_{1ni} & \cdots & \underline{b}_{1mi} \end{pmatrix}, \overline{B}_i = \begin{pmatrix} \overline{b}_{11i} & \cdots & \overline{b}_{1mi} \\ \vdots & \ddots & \vdots \\ \overline{b}_{1ni} & \cdots & \overline{b}_{1mi} \end{pmatrix} \qquad [5.12b]$$

An equivalent form of matrices $\hat{A}_i \in [\underline{A}_i, \overline{A}_i]$ and $\hat{B}_i \in [\underline{B}_i, \overline{B}_i]$ defined in [5.3] is:

$$\hat{A}_i = A_{i0} + \sum_{v=1}^{n}\sum_{w=1}^{n} e_v \hat{f}_{vwi}\, e_w^T,\ \left|\hat{f}_{vwi}\right| \leq \delta_{vwi}^a,\quad \delta_{vwi}^a = \frac{1}{2}\left(\overline{a}_{vwi} - \underline{a}_{vwi}\right)$$

[5.13a]

$$\hat{B}_i = B_{i0} + \sum_{v=1}^{n}\sum_{w=1}^{m} e_v \hat{g}_{vwi} h_w^T, \ |\hat{g}_{vwi}| \le \delta_{vwi}^b, \ \delta_{vwi}^b = \frac{1}{2}\left(\overline{b}_{vwi} - \underline{b}_{vwi}\right) \quad [5.13b]$$

with the following definitions:

$$A_{i0} = \frac{1}{2}\left(\underline{A}_i + \overline{A}_i\right), \ \Delta A_i = \frac{1}{2}\left(\overline{A}_i - \underline{A}_i\right)$$
$$B_{i0} = \frac{1}{2}\left(\underline{B}_i + \overline{B}_i\right), \ \Delta B_i = \frac{1}{2}\left(\overline{B}_i - \underline{B}_i\right) \quad [5.14]$$

$$e_v = \begin{pmatrix} 0\ldots & \underset{v^{th}\ element}{1} & \ldots 0 \end{pmatrix}^T, \quad e_v \in \mathrm{R}^n \quad [5.15]$$

$$h_w = \begin{pmatrix} 0\ldots & \underset{w^{th}\ element}{1} & \ldots 0 \end{pmatrix}^T, \quad h_w \in \mathrm{R}^m \quad [5.16]$$

For the following, we define:

$$V_{ai} = diag\left(\lambda_{11i}\ldots\lambda_{1ni}\ldots\lambda_{n1i}\ldots\lambda_{nni}\right) > 0 \quad [5.17]$$

$$V_{bi} = diag\left(v_{11i}\ldots v_{1mi}\ldots v_{n1i}\ldots v_{nmi}\right) > 0 \quad [5.18]$$

5.2.2.1. *Stability analysis*

First, we analyze the stability of the open-loop multiple model in its equivalent form:

$$\dot{x}(t) = \sum_{i=1}^{N} \mu_i(z(t)) \left(A_{i0} + \sum_{v=1}^{n}\sum_{w=1}^{n} e_v \hat{f}_{vwi} e_w^T \right) x(t) \quad [5.19]$$

This multiple model is asymptotically stable if a matrix $P > 0$ exists such that:

$$P\hat{A}_i^T + \hat{A}_i P < 0 \quad \forall i \in I_N \qquad [5.20]$$

Based on the conditions above and on Lemma 2.2, the following result proposes robust stability conditions in an LMI form.

THEOREM 5.2.– Let us assume that there exist matrices $P > 0$ and $V_{ai} = diag(\lambda_{11i} \ldots \lambda_{1ni} \ldots \lambda_{n1i} \ldots \lambda_{nni}) > 0$ which satisfy the following LMI $\forall i \in I_N$:

$$\begin{pmatrix} PA_{i0}^T + A_{i0}P + E\Delta_i^a FV_{ai}F\Delta_i^a E^T & PE \\ * & -V_{ai} \end{pmatrix} < 0 \qquad [5.21]$$

where:

$$\begin{aligned} &\Delta_i^a = diag\left(\delta_{11i}^a \ldots \delta_{n1i}^a \ \ \delta_{12i}^a \ldots \delta_{n2i}^a \ \ \delta_{1ni}^a \ldots \delta_{nni}^a\right) \\ &E = \underbrace{\left(\mathrm{I}_n \ldots \mathrm{I}_n\right)}_{n \ \ times} \end{aligned} \qquad [5.22]$$

and F is a permutation matrix such that:

$$FV_{ai}F = diag(\lambda_{11i} \ldots \lambda_{n1i} \ \lambda_{12i} \ldots \lambda_{n2i} \ \lambda_{1ni} \ldots \lambda_{nni}) \qquad [5.23]$$

then multiple model [5.19] is asymptotically stable.

PROOF.– Taking [5.13a] into account, we get:

$$P\hat{A}_i^T + \hat{A}_i P = PA_{i0}^T + A_{i0}P + \sum_{v=1}^{n}\sum_{w=1}^{n}\left(P\left(e_v \hat{f}_{vwi} e_w^T\right)^T + \left(e_v \hat{f}_{vwi} e_w^T\right)P \right)$$

[5.24]

Using Lemma 2.2, equation [5.24] is overvalued by:

$$P\hat{A}_i^T + \hat{A}_i P \leq P A_{i0}^T + A_{i0} P + \sum_{v=1}^{n}\sum_{w=1}^{n}\left(\lambda_{vwi} (\delta_{vwi}^a)^2 e_v e_v^T + \frac{1}{\lambda_{vwi}} \left(P e_w e_w^T P \right) \right) \qquad [5.25]$$

where $\lambda_{vwi} > 0$. We observe that:

$$\sum_{v=1}^{n}\sum_{w=1}^{n} \frac{1}{\lambda_{vwi}} \left(P e_w e_w^T P \right) = \left(P \ldots P \right) V_{ai}^{-1} \left(P \ldots P \right)^T \qquad [5.26a]$$

$$\left(P \ldots P \right) = PE \qquad [5.26b]$$

$$\sum_{v=1}^{n}\sum_{w=1}^{n} \lambda_{vwi} (\delta_{vwi}^a)^2 e_v e_v^T = E \Delta_i^a F V_{ai} F \Delta_i^a E^T \qquad [5.26c]$$

Then conditions [5.21] guarantee $P\hat{A}_i^T + \hat{A}_i P < 0 \quad \forall i \in I_N$.

5.2.2.2. *Control law synthesis*

Let us consider the state feedback control law used in earlier chapters:

$$u(t) = \sum_{i=1}^{N} \mu_i(z(t)) K_i x(t) \qquad [5.27]$$

The closed-loop multiple model becomes:

$$\dot{x}(t) = \sum_{i=1}^{N}\sum_{j=1}^{N} \mu_i(z(t)) \mu_j(z(t)) \left(\hat{A}_i + \hat{B}_i K_j \right) x(t) \qquad [5.28]$$

Consequently, the system represented by [5.29] is asymptotically stable if matrices $P > 0$ and K_i exist such that the following conditions are verified $\forall (i,j) \in I_N^2$:

$$P\left(\hat{A}_i + \hat{B}_i K_j\right)^T + \left(\hat{A}_i + \hat{B}_i K_j\right)P < 0 \qquad [5.29]$$

The following result proposes LMI conditions, which are easy to implement for these hard-to-resolve conditions.

THEOREM 5.3.– If there exist matrices $P > 0$, $V_{ai} > 0$, $V_{bi} > 0$ and Y_i which satisfy the following LMI $\forall (i,j) \in I_N^2$:

$$\Lambda_{ij} = \begin{pmatrix} \Psi_{ij} & * & * \\ E_1^T P & -V_{ai} & * \\ E_2^T Y_j & 0 & -V_{bi} \end{pmatrix} < 0 \qquad [5.30]$$

where:

$$\begin{aligned}\Psi_{ij} &= PA_{i0}^T + A_{i0}P + B_{i0}Y_j + Y_j^T B_{i0}^T \\ &+ E_1 \Delta_i^a F_1 V_1 F_1 \Delta_i^a E_1^T + E_2 \Delta_i^b F_2 V_2 F_2 \Delta_i^b E_2^T\end{aligned} \qquad [5.31a]$$

$$\Delta_i^a = diag\left(\delta_{11i}^a \ldots \delta_{n1i}^a \;\; \delta_{12i}^a \ldots \delta_{n2i}^a \;\; \delta_{1ni}^a \ldots \delta_{nni}^a\right),\; E_1 = \underbrace{\left(\mathrm{I}_n \ldots \mathrm{I}_n\right)}_{n \;\; times} \qquad [5.31b]$$

$$\Delta_i^b = diag\left(\delta_{11i}^b \ldots \delta_{n1i}^b \;\; \delta_{12i}^b \ldots \delta_{n2i}^b \ldots \delta_{1mi}^b \ldots \delta_{nmi}^b\right), E_2 = \underbrace{\left(\mathrm{I}_m \ldots \mathrm{I}_m\right)}_{n \;\; times} \qquad [5.31c]$$

and F_1, F_2 are permutation matrices such that:

$$F_1 V_{ai} F_1 = diag\left(\lambda_{11i} \ldots \lambda_{n1i} \;\; \lambda_{12i} \ldots \lambda_{n2i} \;\; \lambda_{1ni} \ldots \lambda_{nni}\right) \qquad [5.32a]$$

$$F_2 V_{bi} F_2 = diag\left(\nu_{11i} \ldots \nu_{n1i} \;\; \nu_{12i} \ldots \nu_{n2i} \;\; \ldots \;\; \nu_{1mi} \ldots \nu_{nmi}\right) \qquad [5.32b]$$

then multiple model [5.29] with control law [5.28] and $K_i = Y_i P^{-1}$ is asymptotically stable.

Some relaxations can be carried out on [5.30] by considering the techniques used in earlier chapters. Indeed the conditions $\Lambda_{ij} < 0$ defined in [5.30] can be replaced by the less conservative conditions of Lemma 3.1 $\forall i, j \in I_N$ as follows:

$$P > 0,\ \Lambda_{ii} + Q_{ii} < 0 \qquad [5.33a]$$

$$\Lambda_{ij} + \Lambda_{ji} + Q_{ij} + Q_{ij}^T \leq 0 \qquad [5.33b]$$

$$Q = Q^T = \begin{pmatrix} Q_{11} & \cdots & Q_{1N} \\ * & \ddots & \vdots \\ * & * & Q_{NN} \end{pmatrix} > 0 \qquad [5.33c]$$

In the case of a common control matrix with no uncertainties $B_i = B_{i0}$, LMI [5.30]–[5.32] are expressed $\forall (i, j) \in I_N^2$ as:

$$\begin{pmatrix} PA_{i0}^T + A_{i0}P + B_{i0}Y_j + Y_j^T B_{i0}^T + E_1 \Delta_i^a F_1 V_1 F_1 \Delta_i^a E_1^T & E_1 P \\ * & -V_{ai} \end{pmatrix} < 0$$

[5.34]

EXAMPLE 5.1.– Let us consider the following example:

$$A_{10} = \begin{pmatrix} -1 & -1.155 \\ 1 & 0 \end{pmatrix},\quad \Delta A_1 = \begin{pmatrix} 0 & 6.55 \\ 0 & 0 \end{pmatrix}$$

$$A_{20} = \begin{pmatrix} -1 & -1.155 \\ 1 & 0 \end{pmatrix},\quad \Delta A_2 = \begin{pmatrix} 0 & 6.55 \\ 0 & 0 \end{pmatrix}$$

$$B_{10} = \begin{pmatrix} 1.4687 \\ 0 \end{pmatrix},\ B_{20} = \begin{pmatrix} 0.5613 \\ 0 \end{pmatrix},\ \Delta B_1 = \Delta B_2 = \begin{pmatrix} 0 \\ 0 \end{pmatrix}$$

Conditions [5.32] and [5.33] which correspond to $B_i = B_{i0}$, where $n=2$, $E_1 = (I_2\ I_2)$, $V_{ai} = diag(\lambda_{11}, \lambda_{12}, \lambda_{21}, \lambda_{22})$, can be solved with:

$$P = \begin{pmatrix} 6.61 & -2.49 \\ -2.49 & 1.63 \end{pmatrix},\ V_{ai} = (8.75, 8.75, 8.75, 8.75)$$

The control law gains are:

$K_1 = (127.91\ 197.04)$, $K_2 = (127.91\ 197.04)$

5.3. Output feedback control

The output feedback control of physical systems, and in particular those described by uncertain multiple models, is one of the major challenges that is hard to solve. We first consider state output feedback control by considering the two types of uncertainties: structured norm-bounded uncertainties and interval uncertainties.

5.3.1. *Norm-bounded uncertainties*

The uncertain multiple model [5.1] is now considered with the following hypothesis:

HYPOTHESIS 5.1.– Output matrix $C_i = C, i:1,...,N$ is with full rank (i.e. the rank is equal to the number of rows of C).

The output feedback control is of the form:

$$u(t) = \sum_{i=1}^{N} \mu_i(z(t)) K_i y(t) \qquad [5.35]$$

where $K_i \in \mathbb{R}^{m \cdot l}$ are the control law parameters to be determined. The closed-loop multiple model is written as:

$$\dot{x}(t) = \sum_{i=1}^{N}\sum_{j=1}^{N} \mu_i(z(t))\mu_j(z(t))\left(A_i + \Delta A_i + \left(B_i + \Delta B_i\right)K_j C\right)x(t) \quad [5.36]$$

Using the relaxed conditions given above (Lemma 3.1), this model is asymptotically stable if positive definite symmetric matrices Q_{ii}, P and Q_{ij} exist that verify the following conditions $\forall (i,j) \in I_N^2, i < j$:

$$P > 0 \quad [5.37a]$$

$$P\left(A_i + \Delta A_i + \left(B_i + \Delta B_i\right)K_j C\right)^T + \left(A_i + \Delta A_i + \left(B_i + \Delta B_i\right)K_j C\right)P + Q_{ii} < 0 \quad [5.37b]$$

$$P\left(A_i + A_j + \Delta A_j + \Delta A_i + \left(B_i + \Delta B_i\right)K_j C + \left(B_j + \Delta B_j\right)K_i C\right)^T + \left(A_i + A_j + \Delta A_j + \Delta A_i + \left(B_i + \Delta B_i\right)K_j C + \left(B_j + \Delta B_j\right)K_i C\right)P + Q_{ij} + Q_{ij}^T \leq 0 \quad [5.37c]$$

$$Q = Q^T = \begin{pmatrix} Q_{11} & \cdots & Q_{1N} \\ * & \ddots & \vdots \\ * & * & Q_{NN} \end{pmatrix} > 0 \quad [5.37d]$$

These nonlinear conditions are transformed into linear constraints by applying the techniques used above and Lemma 2.2.

THEOREM 5.4.– If there exist matrices N_i, M, Q_{ij} and $P>0$ and positive scalars ε_{ij}, κ_{ij} such that $\forall i<j\in I_N$:

$$\Sigma_{ii} = \begin{pmatrix} T_{ii}+Q_{ii} & * & * \\ E_{Ai}P & -\varepsilon_{ii}I & * \\ E_{Bi}N_iC & 0 & -\kappa_{ii}I \end{pmatrix} < 0 \qquad [5.38a]$$

$$\Sigma_{ij} = \begin{pmatrix} T_{ij}+T_{ji}+Q_{ij}+Q_{ij}^T & * & * \\ \begin{pmatrix} E_{Ai} \\ E_{Aj} \end{pmatrix} P & -\begin{pmatrix} \varepsilon_{ij}I & 0 \\ 0 & \varepsilon_{ji}I \end{pmatrix} & * \\ \begin{pmatrix} E_{Bi}N_j \\ E_{Bj}N_i \end{pmatrix} C & 0 & -\begin{pmatrix} \delta_{ij}I & 0 \\ 0 & \delta_{ji}I \end{pmatrix} \end{pmatrix} < 0$$

[5.38b]

$$Q>0 \qquad [5.38c]$$

$$CP = MC \qquad [5.38d]$$

$$T_{ij} = PA_i^T + A_iP + C^TN_j^TB_i^T + B_iN_jC + \varepsilon_{ij}D_iD_i^T + \delta_{ij}D_iD_i^T$$

[5.39]

with Q defined in [5.36d], then the multiple model [5.36], with control law $u(t)=\sum_{i=1}^{N}\mu_i(z(t))K_iy(t)$ where $K_i = N_iM^{-1}$, is asymptotically stable.

Without loss of generality, we assume that matrix C has a specific structure of the form $C=(I, 0)$, where I is the identity matrix and 0 is the null matrix with appropriate dimensions. Indeed, since C is with full rank (Hypothesis 5.1), then there always exists a nonsingular transformation matrix T such that $CT=(I,0)$. Indeed, let $x(t)=T\varsigma(t)$, then multiple model [5.1] becomes:

$$\begin{aligned}&\dot{\varsigma}(t)=\sum_{i=1}^{N}\mu_i(z(t))\left(T^{-1}\hat{A}_iT\varsigma(t)+T^{-1}\hat{B}_iu(t)\right)\\&y(t)=CT\varsigma(t)=(I,\,0)\varsigma(t)\end{aligned} \quad [5.40]$$

COROLLARY 5.1.– If there exist matrices N_i, S_{ij} and $P=diag(P_1,P_2)>0$ and positive scalar ε_{ij}, κ_{ij} such that the following LMI are satisfied $\forall i<j\in I_N$:

$$\Sigma_{ii}<0 \quad [5.41a]$$

$$\Sigma_{ij}<0 \quad [5.41b]$$

$$Q>0 \quad [5.41c]$$

with Σ_{ii},Σ_{ij} defined in [5.38], then multiple model [5.36], with control law $u(t)=\sum_{i=1}^{N}\mu_i(z(t))K_iy(t)$ where $K_i=N_iP^{-1}$, is asymptotically stable.

We note that this result enables us to introduce multiple matrices [5.41c]. Consequently, this result leads to a greater freedom in the synthesis of the stabilizing control law (see Lemma 3.1). These strict conditions are also formulated in an LMI form and so are easily solved using existing numeric tools.

EXAMPLE 5.2.– The vehicle model with two degrees of freedom (see Example 1.1) is represented in Figure 5.1, and its differential equation is as follows:

$$\begin{pmatrix} \dot{\beta}(t) \\ \dot{r}(t) \end{pmatrix} = \begin{pmatrix} 2\dfrac{F_f(t)+F_r(t)}{mU} - r(t) \\ 2\dfrac{a_f F_f(t) - a_r F_r(t)}{I_z} \end{pmatrix} \quad [5.42]$$

where β represents the lateral slip angle, r is the yaw rate angle, F_f is the cornering force of the two front tires, F_r is the cornering force of the two rear tires, U is the vehicle speed, which is assumed to be constant, I_z is the moment of inertia relative to the z-axis and m is the mass of the vehicle.

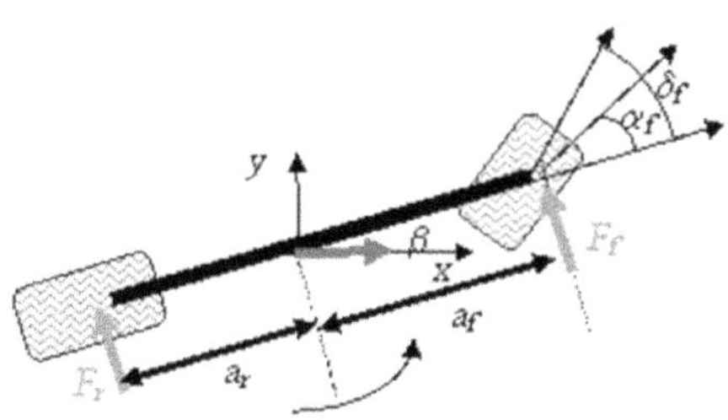

Figure 5.1. *Vehicle model*

The vehicle parameters are given in Table 5.1.

Parameters	I_z	m	a_f	a_r	U
Values	3,000 (kg m²)	1,500 (kg)	1.3 (m)	1.2 (m)	20 (m/s)

Table 5.1. *Vehicle parameters*

The cornering forces (F_f and F_r) have nonlinear behavior (see Figure 1.1). They are dependent on tire slip angles

$\alpha_i\left(i=f,r\right)$ and adhesion coefficient μ. These forces can be approximated using a multiple model:

$$\begin{cases} F_f = h_1(|\alpha_f|)C_{f1}(\mu)\alpha_f + h_2(|\alpha_f|)C_{f2}(\mu)\alpha_f \\ F_r = h_1(|\alpha_f|)C_{r1}(\mu)\alpha_r + h_2(|\alpha_f|)C_{r2}(\mu)\alpha_r \end{cases} \quad [5.43]$$

where $h_i(\cdot)$ $(i=1,2)$ are activation functions such that $h_1(\cdot)+h_2(\cdot)=1$ with the parameters given in Table 5.1.

Under small angle hypothesis, the front and rear slip angles can be approximated by:

$$\alpha_f \cong -\beta - \frac{a_f r}{U} + \delta,\ \alpha_r \cong -\beta + \frac{a_r r}{U} \quad [5.44]$$

where δ is the steering angle. Consequently, nonlinear model [5.42] can be represented by the following multiple model:

$$\begin{cases} \dot{x}(t) = \sum\limits_{i=1}^{2} h_i(|\alpha_f|)\left(A_i x(t) + B_{fi}\delta(t)\right) \\ y(t) = Cx(t) \end{cases} \quad [5.45]$$

where $x(t)=\left(\beta \quad r\right)^T$ and

$$A_i = \begin{pmatrix} -2\dfrac{C_{fi}+C_{ri}}{mU} & -2\dfrac{C_{fi}a_f - C_{ri}a_r}{mU^2} - 1 \\ -2\dfrac{C_{fi}a_f - C_{ri}a_r}{I_z} & -2\dfrac{C_{fi}a_f^2 + C_{ri}a_r^2}{I_z U} \end{pmatrix} \quad [5.46]$$

$$B_{fi}=\begin{pmatrix}\dfrac{2C_{fi}}{mU}\\ \dfrac{2a_fC_{fi}}{I_z}\end{pmatrix},C=\begin{pmatrix}0 & 1\end{pmatrix} \quad [5.47]$$

In order to take into account the variations in coefficients C_{fi} and C_{ri}, which are dependent on the adhesion coefficient, these coefficients are assumed varying as follows:

$$\begin{cases}C_{fi}=C_{fi0}(1+d_if_i)\\ C_{ri}=C_{ri0}(1+d_if_i)\end{cases} \quad [5.48]$$

where $|f_i|\leq 1$ and d_i represents the deviation of the coefficient from its nominal value. The following uncertain multiple model is therefore obtained:

$$\begin{cases}\dot{x}(t)=\sum\limits_{i=1}^{2}h_i(|\alpha_f|)\Big(\big(A_{i0}+\Delta A_i\big)x(t)+\big(B_{fi0}+\Delta B_{fi}\big)\delta(t)\Big)\\ y(t)=Cx(t)\end{cases} \quad [5.49]$$

where:

$$A_{i0}=\begin{pmatrix}-2\dfrac{C_{fi0}+C_{ri0}}{mU} & -2\dfrac{C_{fi0}a_f-C_{ri0}a_r}{mU^2}-1\\ -2\dfrac{C_{fi0}a_f-C_{ri0}a_r}{I_z} & -2\dfrac{C_{fi0}a_f^2+C_{ri0}a_r^2}{I_zU}\end{pmatrix} \quad [5.50]$$

$$B_{fi0}=\begin{pmatrix}\dfrac{2C_{fi0}}{mv}\\ \dfrac{2a_fC_{fi0}}{I_z}\end{pmatrix},\ C=\begin{pmatrix}0 & 1\end{pmatrix} \quad [5.51]$$

$$\Delta A_i = D_i F_i E_{Ai}, \Delta B_{fi} = D_i F_i E_{fBi}, \Delta B_{ri} = D_i F_i E_{rBi} \quad [5.52]$$

$$D_i = \begin{pmatrix} d_i & 0 \\ 0 & d_i \end{pmatrix}, \; F_i = \begin{pmatrix} f_i & 0 \\ 0 & f_i \end{pmatrix} \quad [5.53]$$

$$E_{fBi} = \begin{pmatrix} \dfrac{2C_{fi0}}{mv} \\ \dfrac{2a_f C_{fi0}}{I_z} \end{pmatrix}, \; E_{rBi} = \begin{pmatrix} \dfrac{2C_{ri0}}{mv} \\ -\dfrac{2a_r C_{ri0}}{I_z} \end{pmatrix} \quad [5.54]$$

$$E_{Ai} = \begin{pmatrix} -2\dfrac{C_{fi0}+C_{ri0}}{mU} & -2\dfrac{C_{fi0}a_f - C_{ri0}a_r}{mU^2} \\ -2\dfrac{C_{fi0}a_f - C_{ri0}a_r}{I_z} & -2\dfrac{C_{fi0}a_f^2 + C_{ri0}a_r^2}{I_z U} \end{pmatrix} \quad [5.55]$$

Our aim is to synthesize a robust control law of the form $\delta(t) = \sum_{i=1}^{2} h_i(z(t)) F_i y(t)$. For this, the following data are considered:

$$C_{f10} = 60{,}712, \; C_{f20} = 4{,}812, \; C_{r10} = 60{,}088, \; C_{r20} = 3{,}455 \quad [5.56]$$

$$D_1 = D_2 = \begin{pmatrix} 0.4 & 0 \\ 0 & 0.4 \end{pmatrix} \quad [5.57]$$

The solution to the conditions of Theorem 5.4 gives:

$$K_1 = -2.5477, \; K_2 = -14.7073 \quad [5.58]$$

Simulations are carried out by considering the profile from Figure 5.2 and the adhesion coefficient variations from

Figure 5.3. The simulation results (Figures 5.4 and 5.5) show the robustness of the synthesis conditions.

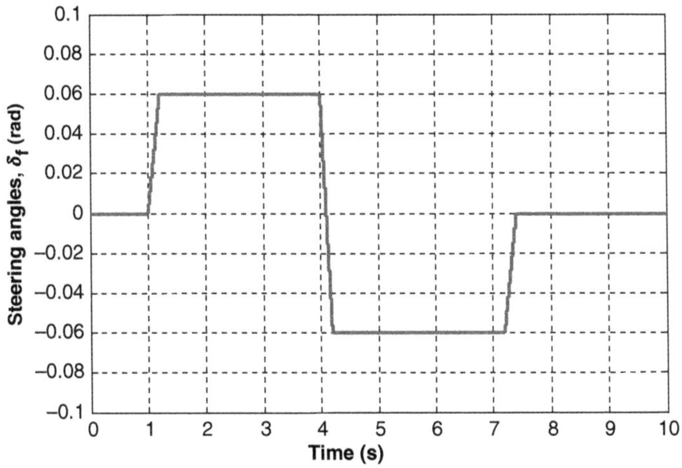

Figure 5.2. *Steering angle profile (rad)*

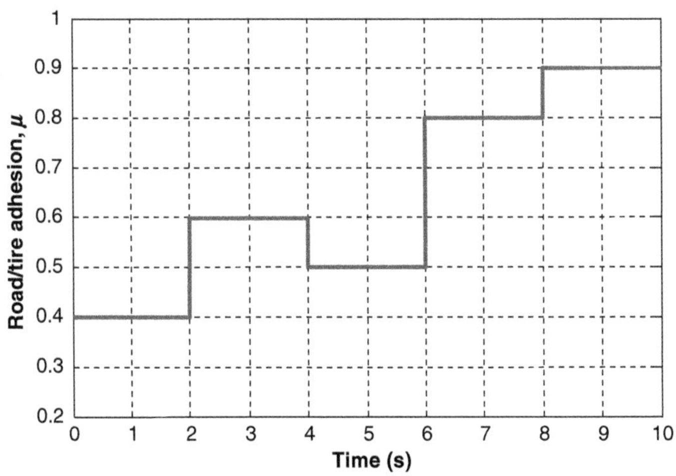

Figure 5.3. *Variation of adhesion coefficient*

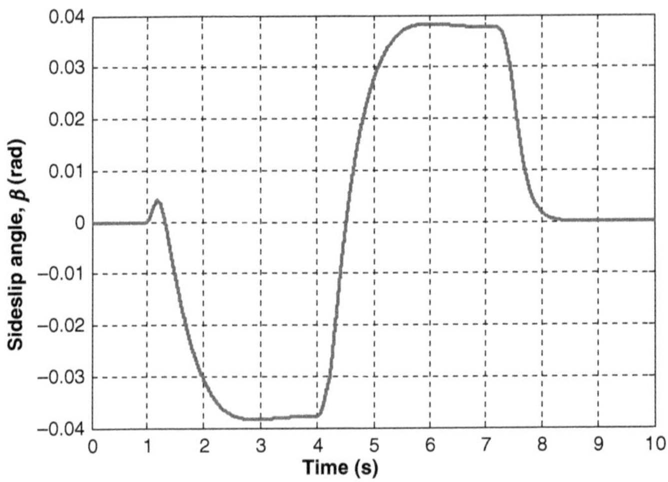

Figure 5.4. *Evolution of $\beta(t)$*

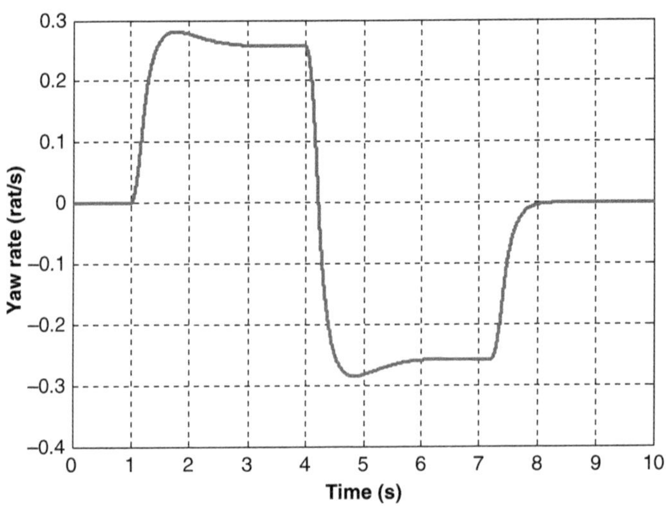

Figure 5.5. *Evolution of $r(t)$*

5.3.2. *Interval uncertainties*

The evolution of the closed-loop multiple model, via control law [5.35], is described by the following equation:

$$\dot{x}(t) = \sum_{i=1}^{N}\sum_{j=1}^{N} \mu_i(z(t))\mu_j(z(t))\left(\hat{A}_i + \hat{B}_i K_j C\right)x(t) \qquad [5.59]$$

This model is globally asymptotically stable [TAN 98]:

$$P\left(\hat{A}_i + \hat{B}_i K_i C\right)^T + \left(\hat{A}_i + \hat{B}_i K_i C\right)P + (r-1)S < 0 \qquad [5.60]$$

$$P\left(\hat{A}_i + \hat{B}_i K_j C + \hat{A}_j + \hat{B}_j K_i C\right)^T + \left(\hat{A}_i + \hat{B}_i K_j C + \hat{A}_j + \hat{B}_j K_i C\right)P - 2S \le 0 \qquad [5.61]$$

The following result proposes strict conditions in an LMI form [CHA 07].

THEOREM 5.5.– Let us assume that there exist matrices $P = diag(P_1, P_2) > 0$, $S \ge 0$, M, N_i and diagonal matrices V_{ai}, V_{bi} which satisfy the following LMI $\forall i < j \in I_N$:

$$\begin{pmatrix} \Phi_{ii} + (r-1)S & PE_1 & C^T N_i^T E_2 \\ * & -V_{ai} & 0 \\ * & * & -V_{bi} \end{pmatrix} < 0 \qquad [5.62]$$

$$\begin{pmatrix} \Phi_{ij} + \Phi_{ji} + 2S & PE_1 & PE_1 & C^T N_j^T E_2 & C^T N_j^T E_2 \\ * & -V_{ai} & 0 & 0 & 0 \\ * & * & -V_{aj} & 0 & 0 \\ * & * & * & -V_{bi} & 0 \\ * & * & * & * & -V_{bj} \end{pmatrix} < 0 \qquad [5.63]$$

where $F_1, F_2, \Delta_i^a, \Delta_i^b$ and Φ_{ij} are defined in [5.31], then multiple model [5.59] derived from its control law $u(t) = \sum_{i=1}^{N} \mu_i(z(t)) K_i y(t)$ where $K_i = N_i M^{-1}$ is asymptotically stable.

It is interesting to note that for a common input matrix $B_i = B$, multiple model [5.59] is globally asymptotically stable if there exist matrices $P = diag(P_1, P_2) > 0$, $S \geq 0$, M, N_i, and diagonal matrices V_{ai}, V_{bi} that satisfy the following $\forall i \in I_N$:

$$\begin{pmatrix} PA_{i0}^T + A_{i0}P + B_{i0}N_iC + (B_{i0}N_iC)^T + E_1\Delta_i^a F_1 V_1 F_1 \Delta_i^a E_1^T & E_1 P \\ * & -V_{ai} \end{pmatrix} < 0 \quad [5.64]$$

where E_1, F_1, Δ_i^a are defined in [5.31] and the controller is determined by $K_i = N_i M^{-1}$.

We observe that the number of LMI is reduced to n instead of n^2 for the general case.

In the case of a linear output feedback $u(t) = Ky(t)$, the synthesis conditions of Theorem 5.5 is simplified with $N_i = N$.

EXAMPLE 5.3.– Let us consider the following uncertain multiple model:

$$\overline{A}_1 = \begin{pmatrix} -4.8 & -1 \\ 2.5 & -3.75 \end{pmatrix}, \; \underline{A}_1 = \begin{pmatrix} -6.72 & -1.4 \\ 1.5 & -5.25 \end{pmatrix} \quad [5.65a]$$

$$\overline{A}_2 = \begin{pmatrix} -0.64 & -1 \\ -1 & -0.5 \end{pmatrix}, \; \underline{A}_2 = \begin{pmatrix} -0.96 & -1.5 \\ 1.5 & -0.75 \end{pmatrix} \quad [5.65b]$$

$$\overline{B}_1 = \begin{pmatrix} 3 \\ -21.2 \end{pmatrix}, \underline{B}_1 = \begin{pmatrix} 2 \\ -31.22 \end{pmatrix} \quad [5.65c]$$

$$\overline{B}_2 = \begin{pmatrix} 0.3 \\ -3.4 \end{pmatrix}, \underline{B}_2 = \begin{pmatrix} 0.2 \\ -3.4 \end{pmatrix} \quad [5.65d]$$

$$C = \begin{pmatrix} 1 & 0 \end{pmatrix} \quad [5.65e]$$

The stability conditions of Theorem 5.6 lead to the resolution of the three LMI, which gives:

$$P = \begin{pmatrix} 3.24 & 0 \\ 0 & 4.66 \end{pmatrix} \quad [5.66]$$

$$V_{a1} = diag(7.3\ 7.3\ 7.3\ 7.3) \quad [5.67a]$$

$$V_{a2} = diag(19.6\ 19.6\ 19.6\ 19.6) \quad [5.67b]$$

$$V_{b1} = diag(0.088\ 0.088) \quad [5.67c]$$

$$V_{b2} = diag(12.69\ 12.69) \quad [5.67d]$$

$$Q_{11} = \begin{pmatrix} 14.24 & 9.7 \\ 9.7 & 24.04 \end{pmatrix}, Q_{22} = \begin{pmatrix} 2.47 & 1.72 \\ 1.72 & 1.31 \end{pmatrix} \quad [5.68]$$

$$Q_{12} = \begin{pmatrix} -4.75 & -3.48 \\ -6.75 & -4.86 \end{pmatrix} \quad [5.69]$$

the controller gains are:

$$K_1 = 0.1094, K_2 = -0.7142 \quad [5.70]$$

Remember that matrix Q_{12}, which is neither symmetric nor positive definite, is the source of relaxations introduced into these conditions.

5.4. Observer-based control

The multiple observer considered is of the form:

$$\begin{aligned}\dot{\hat{x}}(t) &= \sum_{i=1}^{N} \mu_i(z(t))\left(A_i\hat{x}(t) + B_i u(t) + G_i(y(t) - \hat{y}(t))\right)\\ \hat{y}(t) &= \sum_{i=1}^{N} \mu_i(z(t))C_i\hat{x}(t)\end{aligned} \qquad [5.71]$$

where $G_i \in \mathbb{R}^{n.l}$ are the observer gains to be determined. The estimation error is defined by:

$$e(t) = x(t) - \hat{x}(t) \qquad [5.72]$$

The goal is to calculate a reconstructed full state feedback control for multiple model [5.4], in the form:

$$u(t) = \sum_{i=1}^{N} \mu_i(z(t))K_i\hat{x}(t) \qquad [5.73]$$

where $K_i \in \mathbb{R}^{m.n}$ are the control law gains. If [5.71]–[5.73] are taken into account, we obtain:

$$\begin{aligned}\dot{x}(t) = \sum_{i=1}^{N}\sum_{j=1}^{N} \mu_i(z(t))\mu_j(z(t))\\ \left(\left(A_i + \Delta A_i + (B_i + \Delta B_i)K_j\right)x(t) + (B_i + \Delta B_i)K_j e(t)\right)\end{aligned} \qquad [5.74]$$

$$\dot{\hat{x}}(t) = \sum_{i=1}^{N}\sum_{j=1}^{N} \mu_i(z(t))\mu_j(z(t))\left((A_i + B_iK_j)\hat{x}(t) + G_iC_je(t)\right) \qquad [5.75]$$

$$\begin{aligned}\dot{e}(t) &= \sum_{i=1}^{N}\sum_{j=1}^{N}\mu_i(z(t))\mu_j(z(t))\left(A_i - G_iC_j + \Delta B_iK_j\right)e(t) \\ &+ \sum_{i=1}^{N}\sum_{j=1}^{N}\mu_i(z(t))\mu_j(z(t))\left(\Delta A_i + \Delta B_iK_j\right)x(t)\end{aligned} \qquad [5.76]$$

The following result gives sufficient conditions for the observer and controller synthesis stabilizing uncertain multiple model [5.4].

THEOREM 5.6.– If there exist matrices $Q>0$, $P>0$, M_i, N_i and scalars ε_{ij} which satisfy the conditions $\forall i<j\in I_N$:

$$\begin{pmatrix} \Phi_{ii} & * & * \\ E_{Ai}Q+E_{Bi}M_i & -\left(\varepsilon_{ii}^{-1}+1\right)^{-1}I & * \\ D_i^T & 0 & -\left(\varepsilon_{ii}+1\right)^{-1}I \end{pmatrix}<0 \qquad [5.77a]$$

$$\begin{pmatrix} \Psi_{ij} & * & * & * & * \\ E_{Ai}Q+E_{Bi}M_j & -\left(\varepsilon_{ij}^{-1}+1\right)^{-1}I & * & * & * \\ E_{1j}Q+E_{Bj}M_i & 0 & -\left(\varepsilon_{ij}^{-1}+1\right)^{-1}I & * & * \\ D_i^T & 0 & 0 & -\varepsilon_{ij}^{-1} & * \\ D_j^T & 0 & 0 & 0 & -\varepsilon_{ij}^{-1} \end{pmatrix}<0 \qquad [5.77b]$$

$$\begin{pmatrix} T_{ii} & * & * \\ E_{Bi}K_i & -\left(\varepsilon_{ii}^{-1}+1\right)^{-1}I & * \\ D_i^TP & 0 & -\left(\varepsilon_{ii}+1\right)^{-1}I \end{pmatrix}<0 \qquad [5.77c]$$

$$\begin{pmatrix} \Theta_{ij} & * & * & * & * \\ E_{Bi}K_j & -\left(\varepsilon_{ij}^{-1}+1\right)^{-1} I & * & * & * \\ E_{Bj}K_i & 0 & -\left(\varepsilon_{ij}^{-1}+1\right)^{-1} I & * & * \\ D_{ii}{}^T P & 0 & 0 & -\left(\varepsilon_{ij}+1\right)^{-1} I & * \\ D_j{}^T P & 0 & 0 & 0 & -\left(\varepsilon_{ij}+1\right)^{-1} I \end{pmatrix} < 0 \quad [5.77d]$$

where:

$$\Phi_{ii} = QA_i^T + A_iQ + M_i^T B_i^T + B_iM_i + I \quad [5.78a]$$

$$\begin{aligned}\Psi_{ij} = {} & QA_i^T + A_iQ + QA_j^T + A_jQ + M_j^T B_i^T + B_iM_j + M_i^T B_j^T \\ & + B_jM_i + D_iD_i^T + D_jD_j^T + 2I\end{aligned} \quad [5.78b]$$

$$T_{ii} = A_i^T P + PA_i - C_i^T N_i^T - N_iC_i + K_i^T B_i^T B_iK_i \quad [5.78c]$$

$$\begin{aligned}\Theta_{ij} = {} & A_i^T P + PA_i + A_j^T P + PA_j - C_i^T N_j^T - N_iC_j - C_j^T N_i^T \\ & - N_jC_i + K_i^T B_j^T B_jK_i + K_j^T B_i^T B_iK_j\end{aligned} \quad [5.78d]$$

then uncertain multiple model [5.4] is globally asymptotically stable via controller [5.73], based on observer [5.71] with:

$$K_i = M_iQ^{-1} \quad [5.79a]$$

$$G_i = P^{-1}N_i \quad [5.79b]$$

PROOF.– Considering the following candidate Lyapunov function:

$$V\left(x(t), e(t)\right) = V_1\left(x(t)\right) + V_2\left(e(t)\right) \quad [5.80]$$

where:

$$V_1(x(t)) = x(t)^T P_1 x(t), \quad P_1 = Q^{-1} > 0,$$
$$V_2(e(t)) = e(t)^T P e(t), \quad P > 0$$

The time-derivative of $V_1(x(t))$ along the trajectories of [5.74] is:

$$\dot{V}_1(x(t)) = \sum_{i=1}^{N}\sum_{j=1}^{N} \mu_i(z(t))\mu_j(z(t)) \left(x(t)^T \left(H_{ij}^T P_1 + P_1 H_{ij}\right)x(t) + 2x(t)^T P_1 \left(B_i + \Delta B_i\right) K_j e(t)\right) \quad [5.81]$$

where:

$$H_{ij} = A_i + \Delta A_i + \left(B_i + \Delta B_i\right) K_j \quad [5.82]$$

Using Lemma 2.2, the following inequality is obtained:

$$2x(t)^T P_1 \left(B_i + \Delta B_i\right) K_j e(t) \le x(t)^T \left(P_1^2 + P_1 D_i D_i^T P_1\right) x(t) + e(t)^T K_j^T \left(B_i^T B_i + E_{Bi}^T E_{Bi}\right) K_j e(t) \quad [5.83]$$

Taking [5.83] into account in [5.81], the time-derivative of $V_1(x(t))$ is:

$$\dot{V}_1(x(t)) \le \sum_{i=1}^{N}\sum_{j=1}^{N} \mu_i(z(t))\mu_j(z(t)) x(t)^T \left(H_{ij}^T P_1 + P_1 H_{ij} + P_1^2 + P_1 D_i D_i^T P_1\right) x(t) + \sum_{i=1}^{N}\sum_{j=1}^{N} \mu_i(z(t))\mu_j(z(t)) e(t)^T K_j^T \left(B_i^T B_i + E_{Bi}^T E_{Bi}\right) K_j e(t) \quad [5.84]$$

The time-derivative of $V_2(e(t))$ along the trajectories of [5.76] is:

$$\dot{V}_2(e(t)) = \sum_{i=1}^{N}\sum_{j=1}^{N} \mu_i(z(t))\mu_j(z(t)) \left(e(t)^T \left(\Sigma_{ij}^T P + P\Sigma_{ij}\right)e(t) + 2e(t)^T P\left(\Delta A_i + \Delta B_i K_j\right)x(t)\right) \quad [5.85]$$

where:

$$\Sigma_{ij} = A_i - G_i C_j + \Delta B_i K_j \quad [5.86]$$

Using Lemma 2.2 again, we obtain:

$$2e(t)^T P\left(\Delta A_i + \Delta B_i K_j\right)x(t) \le e(t)^T \left(PD_i D_i^T P\right)e(t) + x(t)^T \left(\left(E_{Ai} + E_{Bi}K_j\right)^T \left(E_{Ai} + E_{Bi}K_j\right)\right)x(t) \quad [5.87]$$

The time-derivative of $V_2(e(t))$ becomes:

$$\dot{V}_2(e(t)) \le \sum_{i=1}^{N}\sum_{j=1}^{N} \mu_i(z(t))\mu_j(z(t))e(t)^T \left(\Sigma_{ij}^T P + P\Sigma_{ij} + PD_i D_i^T P\right)e(t) + \sum_{i=1}^{N}\sum_{j=1}^{N} \mu_i(z(t))\mu_j(z(t))x(t)^T \left(\left(E_{Ai} + E_{Bi}K_j\right)^T \left(E_{Ai} + E_{Bi}K_j\right)\right)x(t) \quad [5.88]$$

Consequently, the time-derivative of $V(x(t), e(t))$ can be written as follows:

$$\begin{aligned}
&\dot{V}(x,e) \\
&\leq \sum_{i=1}^{N}\sum_{j=1}^{N}\mu_i(z))\mu_j(z)x(t)^T\left(H_{ij}^T P_1 + P_1 H_{ij} + P_1^2 + P_1 D_i D_i^T P_1 + L_{ij}\right)x(t) \\
&+\sum_{i=1}^{N}\sum_{j=1}^{N}\mu_i(z(t))\mu_j(z(t))e(t)^T\left(\Sigma_{ij}^T P + P\Sigma_{ij} + PD_i D_i^T P + S_{ij}\right)e(t)
\end{aligned} \quad [5.89]$$

where:

$$L_{ij} = \left(E_{Ai} + E_{Bi}K_j\right)^T\left(E_{Ai} + E_{Bi}K_j\right) \quad [5.90]$$

$$S_{ij} = K_j^T\left(B_i^T B_i + E_{Bi}^T E_{Bi}\right)K_j \quad [5.91]$$

which can be rewritten as:

$$\begin{aligned}
\dot{V}(x,e) \leq &\sum_{i=1}^{N}\mu_i(z(t))^2 x(t)^T\left(H_{ii}^T P_1 + P_1 H_{ii} + P_1^2 + P_1 D_i D_i^T P_1 + L_{ii}\right)x(t) \\
&+2\sum_{i<j}^{N}\mu_i(z(t))\mu_j(z(t))x(t)^T\left(\left(\frac{H_{ij}+H_{ji}}{2}\right)^T P_1 + P_1\left(\frac{H_{ij}+H_{ji}}{2}\right)\right)x(t) \\
&+2\sum_{i<j}^{N}\mu_i(z(t))\mu_j(z(t))x(t)^T\left(\frac{2P_1^2 + P_1 D_i D_i^T P_1 + P_1 D_j D_j^T P_1}{2} + \frac{L_{ij}+L_{ji}}{2}\right)x(t) \\
&+\sum_{i=1}^{N}\mu_i(z(t))^2 e(t)^T\left(\Sigma_{ii}^T P + P\Sigma_{ii} + PD_i D_i^T P + S_{ii}\right)e(t) \\
&+2\sum_{i<j}^{N}\mu_i(z(t))\mu_j(z(t))e(t)^T\left(\left(\frac{\Sigma_{ij}+\Sigma_{ji}}{2}\right)^T P + P\left(\frac{\Sigma_{ij}+\Sigma_{ji}}{2}\right)\right. \\
&\left.+\frac{PD_i D_i^T P + PD_j D_j^T P}{2} + \frac{S_{ij}+S_{ji}}{2}\right)e(t)
\end{aligned} \quad [5.92]$$

In order to ensure that $\dot{V}\left(x(t), e(t)\right)$ is negative $\forall \left(x(t)^T, e(t)^T\right) \neq 0$, we can discuss the negativity of each of the five terms in sum [5.92], which enables us to obtain the conditions of the theorem.

Note that the conditions obtained are in BMI form. In order to use the LMI tool, we can sequentially solve these constraints as follows:

– First, calculate variables Q and M_i of LMI [5.77a] and [5.77b].

– Once the gains $K_i, \forall i \in I_N$ are determined from [5.78a], conditions [5.77c] and [5.77d] become linear in P and N_i, which enables us to determine the gains L_i from [5.78b].

Remember that the "separation principle", which consists of separately determining the controller and observer, while guaranteeing the stabilization of the closed-loop system, is not guaranteed.

5.5. Conclusion

In this chapter, we have examined the synthesis of robust control laws for uncertain multiple models, using static output feedback and reconstructed full state feedback. We have considered two types of uncertainties: structured norm-bounded and interval uncertainties. Synthesis conditions are based on the existence of quadratic Lyapunov functions and LMI formulation. This chapter has also shown interest in the introduction of relaxation. Other more sophisticated methods of relaxation, more or less difficult to implement and requiring longer computation time, exist in literature [CHA 12a, GUE 04, TUA 01].

Conclusion

This book is devoted to the analysis and synthesis of control laws for nonlinear systems using the multiple model (or multimodel) approach. These results are exclusively based on Lyapunov's second method and on LMI formalism. Based on quadratic and then nonquadratic Lyapunov functions, some practical results regarding stability and stabilization have been established. Thus, sufficient conditions for stability and stabilization are proposed using the properties of M-matrices. In order to improve synthesis performance (of PDC/OPDC controllers or multiple observers), the pole placement technique in an LMI region (which enables the decay and damping rates, for example, to be fixed) has been tackled.

Static output feedback control laws have been proposed with pole placement in LMI form. Synthesis techniques based on full state feedback control require the whole of the state vector to be available. As this condition is rarely verified, it becomes necessary to use an observer. These observers are constructed by considering modeling uncertainties and with performance constraints.

We have also examined the synthesis of robust control laws, using static output feedback control and observer-based control, for uncertain multiple models. Synthesis conditions are based on the existence of quadratic Lyapunov functions and LMI formulation. Two types of uncertainties have been considered: structured norm-bounded and interval uncertainties.

The results obtained based on quadratic functions are interesting because it is easy to implement a search of a common Lyapunov function and synthesis parameters (controller or observer), which may be formulated as a convex LMI optimization problem and may be solved efficiently. However, the quadratic method turns out to be very conservative due to the ignorance of the information contained in the activation functions. These constraints become still more conservative if the performance constraints from the closed-loop system are added. Examples have been used to illustrate the conservatism of the quadratic method. We have also shown the interest in:

– using other nonquadratic Lyapunov functions of the form $V(x) = \max\left(V_1(x),..,V_N(x)\right)$, where $V_i(x) = x^T P_i x$, $i \in I_N$;

– introducing supplementary variables to reduce the conservatism of these stability conditions.

APPENDICES

Appendix 1

LMI Regions

A1.1. Definition of an LMI region

A subset D of a complex plane is called an nth-order LMI region if there exist a symmetric matrix $\alpha \in \mathbb{R}^{n.n}$ and a matrix $\beta \in \mathbb{R}^{n.n}$ such that:

$$D = \left\{ z \in C : f_D(z) = \alpha + \beta z + \beta^T z < 0 \right\} \qquad \text{[A1.1]}$$

We observe that values for the characteristic function f_D of complex variable z are taken in the set of nth-order Hermitian matrices and that an LMI region is symmetric with the real axis since $f_D^T(z) = f_D(\overline{z})$. This is often verified by regions used for studying the D-stability of a real matrix, since the spectrum of a real matrix is self-adjoint.

THEOREM A1.1.– [CHI 96]: Let $A \in \mathbb{R}^{n.n}$ and D be an LMI region defined by [A1.1]. Matrix A is D-stable if and only if there exists a positive definite symmetric matrix $X \in \mathbb{R}^{n.n}$ such that:

$$M_D(A, X) = \alpha \otimes X + \beta \otimes (AX) + \beta^T \otimes (AX)^T < 0 \qquad \text{[A1.2]}$$

In order to test the D-stability of a real matrix in an intersection of LMI regions, the authors of [CHI 96] have shown that a single positive definite symmetric solution to this LMI is necessary.

Let us consider the region D which results from the intersection of r LMI sub-regions D_i. The eigenvalues of a real matrix belong to D if and only if there exists a positive definite symmetric matrix $X \in \mathbb{R}^{n.n}$ such that:

$$M_{D_i}(A,X) = \alpha_i \otimes X + \beta_i \otimes (AX) + \beta_i^T \otimes (AX)^T < 0 \quad i \in I_r \qquad \text{[A1.3]}$$

which is equivalent to:

$$M_D(A,X) = \alpha \otimes X + \beta \otimes (AX) + \beta^T \otimes (AX)^T = diag\left(M_{D_i}(A,X)\right) < 0 \qquad \text{[A1.4]}$$

where:

$$\alpha = diag(\alpha_1,..,\alpha_r),\ \beta = diag(\beta_1,..,\beta_r) \qquad \text{[A1.5]}$$

This inequality clearly shows that D may itself be formulated as an LMI region. This shows that the LMI approach enables the same matrix X to be considered for all sub-regions of the intersection, while preserving the need for the D-stability condition.

A1.2. Interesting LMI region examples

Inequality [A1.2] can also be written in the following form:

$$\begin{pmatrix} \alpha_{11}X + \beta_{11}AX + \beta_{11}(AX)^T & \cdots & \alpha_{1n}X + \beta_{1n}AX + \beta_{n1}(AX)^T \\ \vdots & \ddots & \vdots \\ \alpha_{n1}X + \beta_{n1}AX + \beta_{1n}(AX)^T & \cdots & \alpha_{nn}X + \beta_{nn}AX + \beta_{nn}(AX)^T \end{pmatrix} < 0 \qquad \text{[A1.6]}$$

where p_{ij} and q_{ij} represent the elements (i, j) of α and β, respectively.

A1.2.1. *Open left half-plane*

$$\Re(z)<0 \Leftrightarrow z+z^T<0 \quad \text{[A1.7]}$$

It is sufficient to take $\alpha = 0$ and $\beta = 1$. The following LMI is deduced from expression [A1.7]:

$$AX+(AX)^T<0 \quad \text{[A1.8]}$$

A1.2.2. *α-stability*

$$\Re(z)<-a \Leftrightarrow 2a+z+z^T<0 \quad \text{[A1.9]}$$

It is sufficient to take $\alpha = 2a$ and $\beta = 1$, which gives the following LMI:

$$2aX+AX+(AX)^T<0 \quad \text{[A1.10]}$$

A1.2.3. *Vertical band*

$$a_1<\Re(z)<a_2 \Leftrightarrow \begin{pmatrix} -2a_2+z+\overline{z} & 0 \\ 0 & 2a_1-(z+\overline{z}) \end{pmatrix}<0 \quad \text{[A1.11]}$$

It is sufficient to take:

$$\alpha=\begin{pmatrix} -2a_2 & 0 \\ 0 & 2a_1 \end{pmatrix}, \beta=\begin{pmatrix} 1 & 0 \\ 0 & -1 \end{pmatrix}$$

which gives the following LMI:

$$\begin{pmatrix} -2a_2X+AX+(AX)^T & 0 \\ 0 & 2a_1-(AX+(AX)^T) \end{pmatrix}<0 \quad \text{[A1.12]}$$

A1.2.4. *Horizontal band*

$$|\Im m(z)| < a \Leftrightarrow \begin{pmatrix} -2a & z-\overline{z} \\ -z+\overline{z} & -2a \end{pmatrix} < 0 \qquad \text{[A1.13]}$$

It is sufficient to take:

$$\alpha = \begin{pmatrix} -2a & 0 \\ 0 & 2a \end{pmatrix}, \beta = \begin{pmatrix} 0 & 1 \\ -1 & 0 \end{pmatrix}$$

which gives the following LMI:

$$\begin{pmatrix} -2aX & AX-(AX)^T \\ -AX+(AX)^T & 2aX \end{pmatrix} < 0 \qquad \text{[A1.14]}$$

A1.2.5. *Disk of radius R, centered at (q,0)*

$$|z-q| < R \Leftrightarrow \begin{pmatrix} -R & z-q \\ \overline{z}-q & -R \end{pmatrix} < 0 \qquad \text{[A1.15]}$$

It is sufficient to take:

$$\alpha = \begin{pmatrix} -R & -q \\ -q & -R \end{pmatrix}, \beta = \begin{pmatrix} 0 & 1 \\ 0 & 0 \end{pmatrix}$$

which gives the following LMI:

$$\begin{pmatrix} -RX & -qX+AX \\ -qX+(AX)^T & -RX \end{pmatrix} < 0 \qquad \text{[A1.16]}$$

In the case where $R = 1$ and $q = 0$ (which corresponds to the discrete case), we obtain:

$$A^T XA - X < 0$$

A1.2.6. *Conical sector*

$$a\Re(z)+\left|b.\Im m(z)\right|<0 \Leftrightarrow \begin{pmatrix} a(z+\overline{z}) & -b(z-\overline{z}) \\ b(z-\overline{z}) & a(z+\overline{z}) \end{pmatrix}<0 \qquad [A1.17]$$

It is sufficient to take:

$$\alpha=\begin{pmatrix} 0 & 0 \\ 0 & 0 \end{pmatrix},\ \beta=\begin{pmatrix} a & -b \\ b & a \end{pmatrix}$$

which gives the following LMI:

$$\begin{pmatrix} a\left(AX+(AX)^T\right) & -b\left(AX-(AX)^T\right) \\ b\left(AX-(AX)^T\right) & a\left(AX+(AX)^T\right) \end{pmatrix}<0 \qquad [A1.18]$$

We have:

$$0<\theta<\frac{\pi}{2},\ \cos(\theta)=\frac{-b}{\sqrt{a^2+b^2}},\ \sin(\theta)=\frac{a}{\sqrt{a^2+b^2}}$$

Thus:

$$\begin{pmatrix} \sin(\theta)\left(AX+(AX)^T\right) & \cos(\theta)\left(AX-(AX)^T\right) \\ -\cos(\theta)\left(AX-(AX)^T\right) & \sin(\theta)\left(AX+(AX)^T\right) \end{pmatrix}<0 \quad [A1.19]$$

Appendix 2

Properties of M-matrices

These definitions can be found in [FOS 93].

DEFINITION A2.1.– *Z*-matrix

A matrix $A = (a_{ij}) \in \mathbb{R}^{p.p}$ for which all off-diagonal terms are negative or null is called a *Z*-matrix.

THEOREM A2.1.– *M*-matrix

In order for a *Z*-matrix *A* to be an *M*-matrix, it is necessary and sufficient that one of the following equivalent properties is verified:

– each eigenvalue of *A* is of a positive real part;

– all of the principal minor determinants of *A* are positive;

– A^{-1} exists and is nonnegative (all of its elements are either positive or null);

– there exists a positive vector v ($v \neq 0$), such that the vector Av is also a strictly positive vector;

$-A$ is a positive quasi-dominant diagonal matrix, i.e. there exists a set of real numbers κ_j such that:

$$\kappa_j a_{jj} - \sum_{i \neq j:1}^{p} \kappa_i \left| a_{ij} \right| > 0, \forall j \in I_p$$

Appendix 3

Stability and Comparison Systems

Study of stability using comparison systems [BOR 74, GRU 76].

A3.1. Vector norms and overvaluing systems

A3.1.1. *Definition of a vector norm*

Let $E = R^n$ be a vector space and $E_1, E_2, \ldots, E_k$ be subspaces of E:

$$E = E_1 \cup E_2 \cup \ldots \cup E_k$$

Let x be a vector of R^n for which the projection in subspace E_i is defined by:

$$x_i = P_i x \qquad [A3.1]$$

Note that $p_i(x) = p(x_i)$ is a scalar norm defined over subspace E_i; it becomes the vector norm:

$$p : R^n \rightarrow R^k, p(x) = \{p_i(x)\}$$

If x and y are two vectors of space E and $\forall i = 1, 2, .., k$, then the following relations are verified:

$$p_i(x_i) \geq 0 \ \forall \ x_i \in E_i$$

$$p_i(x_i) = 0 \Leftrightarrow x_i = 0$$

$$p_i(x_i + y_i) \leq p_i(x_i) + p_i(y_i), \ \forall \ x_i, y_i \in E_i$$

$$p_i(\lambda x_i) = |\lambda| p_i(x_i), \ \forall \ \lambda \in R, \forall \ x_i \in E_i$$

If $k-1$ subspaces of E_i are insufficient to define the whole of space E, the norm vector is said to be surjective. Furthermore, if every two subspaces E_i are disjoint, the vector norm is said to be regular:

$$E_i \cap E_j = (0)_i \cap (0)_j \ \ \forall i \neq j = 1, ..., k$$

A3.1.2. *Definition of a system overvalued from a continuous process*

We define a process for which the evolution into free state is described by the following equation:

$$\dot{x} = A(x,t).a(x,t) = f(x,t)$$
$$x \in R^n, t \in T, \ f : R^n \times T \to R^n$$

The origin is assumed to be the unique equilibrium point of system [A3.1] and this system accepts the following solutions $\forall x_0, t_0 \in T_0, \ T_0 = [t_0, +\infty]$:

$$A : T \times R^n \to R^{n.m}, a : T \times R^n \to R^m$$

DEFINITION A3.1.– The overvaluing system from system [A3.1] relative to the regular vector norm *p(x)* can be defined as follows:

$$D^{+} p(x) \leq M(x,t) p(x) \ \forall x \in E, \forall t \in T_0$$

NOTE:

$$a_j(x,t) = P_{rj} a(x,t), \forall j = 1,2,\ldots,k$$
$$A_{ij}(x,t) = P_{ri} A(x,t) P_{rj}, \forall i,j = 1,2,\ldots,k$$
$$M(x,t) = \{\mu_{ij}(x,t)\}$$

P_{rj} $(j = 1,2,..,k)$ is the projection operator of R^n in ξ_j

$$m_{ij}(x,y,t) = \frac{(grad\ p_i(y_i))^T A_{ij}(x,t) a_i(y_i,t)}{p_i(y_i)} \quad \forall i,j = 1,2,..,k$$

DEFINITION A3.2.– The following matrix:

$$M(x,t) = \{\mu_{ij}(x,t)\}, \ M_{ij} : R^n \times L \to R$$

with:

$$\mu_{ij}(x,t) = \sup_{y \in E}(m_{ii}(x,y,t)\ , i = 1,..,k, x \in E, t \in T_0$$

$$\mu_{ij}(x,t) = \max\left\{0, \sup_{y \in E}(m_{ii}(x,y,t)\right\}, i \neq j = 1,..,k, x \in E, t \in T_0$$

is a pseudo-overvaluing matrix of the process.

DEFINITION A3.3.– The following matrix:

$$M(t) = \{\mu_{ij}(t)\}$$

with:

$$\mu_{ii}(t) = \sup_{y \in E}(m_{ii}(y, y, t)\ , \forall i = 1,.., k, \forall t \in T_0$$

$$\mu_{ij}(t) = \max\left\{0, \sup_{y \in E}(m_{ii}(y, y, t)\right\}, i \neq j = 1,.., k, t \in T_0$$

is a pseudo-overvaluing matrix of the process.

DEFINITION A3.4.– The following matrix:

$$M = \{\mu_{ij}\}$$

with:

$$\mu_{ii} = \sup_{y \in E, t \in T}(m_{ii}(y, y, t))\ , \forall i = 1,.., k$$

$$\mu_{ij} = \max\left\{0, \sup_{y \in E, t \in T}(m_{ij}(y, y, t))\right\}, \forall i \neq j = 1,.., k, y \in E, t \in T_0$$

is a pseudo-overvaluing matrix of the process. This results in the following inequalities:

$$M \geq M(t) \geq M(t, x)$$

If $a(t,x) = x$, it follows that:

$$m_{ij}(y, y, t) = (grad\ p_i(y_i))^T A_{ij}(x, t) y_i\ ,$$

where:

$$A_{ij} = P_{ri} A P_{rj}$$

A3.1.3. *Application*

If the decomposition into subspaces of space E corresponds to a partition into blocks of matrix A, (i, j) represent the indices of the rows and columns of block A_{ij}; if $p_i(x_i)$ is the Euclidean norm of x_i by denoting $a_{ij}(x,t),\ i,j=1,\dots,n$ are the elements of matrix $A(x,t)$. Overvaluing elements are as follows:

$$m_{ii}(x,t)=\max_{k\in I_i}\left(a_{kk}+\frac{1}{2}\sum_{\substack{l\in I_i\\ l\neq k}}\left|a_{kl}+a_{lk}\right|\right),\ i=1,2,..,k$$

$$m_{ij}(x,t)=\frac{1}{2}\left(\max_{k\in I_i}\sum_{l\in I_j}\left|a_{kl}\right|+\max_{l\in I_j}\sum_{k\in I_i}\left|a_{ik}\right|\right),\ \forall i\neq j$$

We observe that if $M(x,t)$ is a pseudo-overvaluing matrix of $A(x,t)$, any matrix $M^*(x,t)=\{m_{ij}^*(x,t)\}$ which verifies $m_{ij}^*(x,t)\geq \mu_{ij}(x,t)$ is also a pseudo-overvaluing matrix of $A(x,t)$.

A3.2. Vector norms and the principle of comparison

We define a system for which the evolution is described by the following equation:

$$\dot{x}=f(x,t)$$

If there exists a vector norm p such that condition:

$$D^+p(x)\leq M(x,t)p(x)$$

is verified:

$$\forall x\in E, \forall t\in T_0$$

with the matrix $M(x,t)$ having its off-diagonal elements nonnegative, then the system for which the evolution is described by the equation:

$$\dot{z} = M(x,t)z,\ z \in R^k$$

is a comparison system of the initial system, and the following condition:

$$z(t_0) \geq p(x,t_0)$$

implies:

$$z(t) \geq p(x,t),\ \forall t > t_0$$

A3.3. Application to stability analysis

For a process, if there exists a stable comparison system, then the same holds true for the initial system. A very significant case results from this: the Borne–Gentina criterion.

If, for an initial system, it is possible to find the initial comparison system such that matrix M can be the opposite of an M-matrix and have its nonconstant elements isolated into a single range, then verification of the linear stability conditions for the comparison system implies stability of the initial system [BOR 87].

Bibliography

[AKH 03] AKENAK A., CHADLI M., RAGOT J., MAQUIN D., "Sliding mode multiple observer for fault detection and isolation", *42nd IEEE Conference on Decision and Control*, vol. 1, pp. 953–958, Hawaii, USA, 9–12 December, 2003.

[AKH 04a] AKENAK A., CHADLI M., RAGOT J., MAQUIN D., "State estimation of uncertain multiple model with unknown inputs", *43rd IEEE Conference on Decision and Control,* Atlantic, Paradise Island, Bahamas, 14–17 December, 2004.

[AKH 04b] AKENAK A., CHADLI M., MAQUIN D., RAGOT J., "Estimation d'état et d'entrées inconnues d'un système non linéaire représenté par un multimodèle", *Conférence Internationale Francophone d'Automatique, CIFA'2004*, Douz, Tunisia, 22–24 November, 2004.

[AKH 07a] AKENAK A., CHADLI M., MAQUIN D., RAGOT J., "Design of a sliding mode fuzzy observer for uncertain Takagi-Sugeno fuzzy model: application to automatic steering of vehicles", *International Journal of Vehicle Autonomous Systems (IJVAS)*, vol. 5, no. 3/4, pp. 288–305, 2007.

[AKH 07b] AKENAK A., CHADLI M., RAGOT J., MAQUIN D., "Multiple model approach modelling: application to a turbojet engine", *Journal of Engineering and Applied Sciences*, vol. 2, no. 4, pp. 798–807, 2007.

[BEN 78] BENREJEB M., BORNE P., "On an algebraic stability criterion for nonlinear processes. Interpretation in the frequency domain", *MECO'78*, pp. 678–682, Athens, Greece, 1978.

[BEN 99] BENNET S.M., PATTON R.J., DALEY S., "Sensor fault-tolerant control of a rail traction drive", *Control Engineering Practice*, vol. 7, no. 2, pp. 217–225, 1999.

[BEN 01] BENABDENNOUR R., BORNE P., KSOURI M., M'SAHLI F., *Identification et commande numérique des procédés industriels*, Technip, Paris, 2001.

[BIE 92] BIEN Z., KIM J.H., "A robust stability bound of linear system with structured uncertainty", *IEEE Transactions on Automatic Control*, vol. 37, no. 10, pp. 1549–1551, 1992.

[BOR 74] BORNE P., GENTINA J.C., "Absolute stability of non-linear, non stationary large scale systems", *Proceedings of Joint Automatic Control Conference*, pp. 317–326, Austin, USA, 1974.

[BOR 87] BORNE P., HAGGEGE J., *Les réseaux de neurones. Présentation et applications*, Technip, Paris, 1987.

[BOR 90] BORNE P., DAUPHIN-TANGUY G., RICHARD J.P., ROTTELLA F., ZAMBETTAKIS I., *Commande et optimisation des processus*, Technip, Paris, 1990.

[BOR 92] BORNE P., DAUPHIN-TANGUY G., RICHARD J.P., ROTTELLA F., ZAMBETTAKIS I., *Modélisation et identification des processus*, Tome 1, Technip, Paris, 1992.

[BOR 93a] BORNE P., "Nonlinear systems stability: vector norm approach", *Systems and Control Encyclopedia*, T5, pp. 3402–3406. 1993.

[BOR 93b] BORNE P., DAUPHIN-TANGUY G., Volumes 1 and 2 in RICHARD J.P., ROTELLA F., ZAMBETTAKIS I. (ed.), *Analyse et régulation des processus industriels*, Technip, Paris, 1993.

[BOR 06] BORNE P., BENREJEB M., HAGGEGE J., "Les réseaux de neurones, Présentation et applications", *Revue d'Electricité et d'Electronique, REE*, no. 9, pp. 37–92, 2006.

[BOR 08] BORNE P., BENREJEB M., "On the representation and the stability study of large scale systems", *IJCC*, vol. 3, pp. 55–66, 2008.

[BOU 00] BOUTAYEB M., "Identification of nonlinear systems in the presence of unknown but bounded disturbances", *IEEE Transactions on Automatic Control*, vol. 45, no. 8, pp. 1503–1507, 2000.

[BOY 94] BOYD S., EL GHAOUI L., FERON E., BALAKRISHNAN V., *Linear Matrix Inequalities in System and Control Theory*, SIAM, Philadelphia, USA, 1994.

[BRE 02] BREGSEN P., PALM R., DRIANKOV D., "Observers for Takagi-Sugeno fuzzy systems", *IEEE Transactions on Systems, Man and Cybernetics*, Part B, vol. 32, no. 1, pp. 114–121, 2002.

[BUC 93] BUCKLEY J.J., "Sugeno type controllers are universal controllers", *Fuzzy Sets and Systems*, vol. 53, no. 3, pp. 299–303, 1993.

[CAS 95] CASTRO J., "Fuzzy logic controllers are universal approximator", *IEEE Transctions on Systems, Man and Cybernetics*, vol. 25, pp. 629–635, 1995.

[CHA 00] CHADLI M., MAQUIN D., RAGOT J., "Relaxed stability conditions for the T-S fuzzy systems", *IEEE International Conference on Systems, Man and Cybernetics*, vol. 5, pp. 3514–3519, 2000.

[CHA 02a] CHADLI M., MAQUIN D., RAGOT J., "Output stabilization in multiple model approach", *International Conference on Control Applications*, vol. 2, pp. 1315–1320, 2002.

[CHA 02b] CHADLI M., Stabilité et stabilisation des multimodèles, thesis, INPL-CRAN, Nancy, 9 December 2002.

[CHA 06a] CHADLI M., ELHAJJAJI A., "Observer-based robust fuzzy control of nonlinear systems with parametric uncertainties-comment on", *Fuzzy Sets and Systems Journal*, vol. 157, no. 9, pp. 1276–1281, 2006.

[CHA 06b] CHADLI M., "On the stability analysis of uncertain fuzzy models", *International Journal of Fuzzy Systems*, vol. 8, no. 4, pp. 224–231, 2006.

[CHA 07a] CHADLI M., BORNE P., "Design of robust static output controller for uncertain fuzzy model: an LMI formulation", *Studies in Informatics and Control*, vol. 16, no. 4, pp. 421–430, 2007.

[CHA 07b] CHADLI M., ELHAJJAJI A., "Moment robust fuzzy observer-based control for improving driving stability", *International Journal of Vehicle Autonomous Systems*, vol. 5, no. 3/4, pp. 326–344, 2007.

[CHA 08a] CHADLI M., AKHENAK A., RAGOT J., MAQUIN D., "On the design of observer for unknown inputs fuzzy models", *International Journal of Automation and Control*, vol. 2, no. 1, pp. 113–125, 2008.

[CHA 08b] CHADLI M., "Multimodèles: origine et méthodes d'obtentions", *Revue d'Electricité et d'Electronique, REE-SEE*, no. 10, pp. 51–54, 2008.

[CHA 08c] CHADLI M., AKHENAK A., MAQUIN D., RAGOT J., "Fuzzy observer for fault detection and reconstruction of unknown input fuzzy models", *International Journal of Modelling, Identification and Control*, vol. 3, no. 2, pp. 193–200, 2008.

[CHA 08d] CHADLI M., BORNE P., "Stabilité des multimodèles", *Revue d'Electricité et d'Electronique, REE-SEE*, no. 10, pp. 55–59, 2008.

[CHA 08e] CHADLI M., "Commande et observation des multimodèles", *Revue d'Electricité et d'Electronique, REE-SEE*, no. 10, pp. 60–65, 2008.

[CHA 09] CHADLI M., AKHENAK A., RAGOT J., MAQUIN D., "State and unknown input estimation for discrete time multiple model", *Journal of the Franklin Institute*, vol. 346, no. 6, pp. 593–610, 2009.

[CHA 10a] CHADLI M., "Chaotic Systems Reconstruction", in *Evolutionary Algorithms and Chaotic Systems*, vol. 267, Springer-Verlag, Berlin, 2010.

[CHA 10b] CHADLI M., "State and an LMI approach to design observer for unknown inputs Takagi-Sugeno Fuzzy models", *Asian Journal of Control*, vol. 12, no. 4, pp.524-530, 2010.

[CHA 12a] CHADLI M., GUERRA T.-M., "LMI solution for robust static output feedback control of Takagi-Sugeno fuzzy models", *IEEE Trans. on Fuzzy Systems*, vol. 20(6), 2012.

[CHA 12b] CHADLI M., KARIMI H.R., "Robust observer design for unknown inputs Takagi-Sugeno models", *IEEE Trans. on Fuzzy Systems*, 2012.

[CHI 96] CHILALI M., GAHINET P., "H∞ Design with pole placement constraints: an LMI approch", *IEEE Transactions on Automatic Control*, vol. 41, no. 3, pp. 358–367, 1996.

[COM 02] COMMAULT C., DION J.M., SENAME O., MOTYEIAN R., "Observer-based fault detection and isolation for structured systems", *IEEE Transactions on Automatic Control*, vol. 47, no. 12, pp. 2074–2079, 2002.

[DAR 94] DAROUACH M., ZASADZINSKI M., XU S.J., "Full-order observers for linear systems with unknown inputs", *IEEE Transactions on Automatic Control*, vol. 39, no. 3, pp. 606–609, 1994.

[DEL 96] DELMOTE F., DUBOIS L., BORNE P., "A general scheme for multi-model controller using trust", *Mathematics and Computers in Simulation*, vol. 41, pp. 173–186, 1996.

[EDW 00] EDWARDS C., SPURGEON S.K., "Sliding mode observers for fault detection and isolation", *Automatica*, vol. 36, no. 4, pp. 541–553, 2000.

[ELF 08] ELFELLY N., DIEULOT J.Y., BORNE P., "A neutral approach of multiple model representation of complex processes", *IJCCC*, vol. 3, no. 2, pp. 39–50, 2008.

[ELF 10a] ELFELLY N., DIEULOT J.Y., BENRJEB M., BORNE P., "A new approach for multiple model identification of complex systems based on both neural and fuzzy clustering algorithms", *Engineering Applications of Artificial Intelligence*, vol. 23, no. 7, pp. 1064–1071, 2010.

[ELF 10b] ELFELLY N., DIEULOT J.Y., BENRJEB M., BORNE P., "A new multiplmodel approach for complex systems modeling based on classification algorithms, experimental validation", *Proceeding of 12th LSS IFAC Symposium*, Villeneuve d'Ascq, France, 2010.

[ELK 99] ELKAMEL A., BORN P., KSOURI L., BENRJEB M., "On the stability of nonlinear multimodel systems", *Inter. J. of Stability and Control: Theory and Applications*, vol. 2. pp. 40-52, 1999.

[FOS 93] FOSSARD A.J., NORMAND-CYROT D., *Système non linéaire - Modélisation-estimation*, tome 1, Masson, Paris, 1993.

[FRA 90] FRANK P.M., "Fault diagnosis in dynamic system using analytical and knowledge based redundancy – a survey and some new results", *Automatica*, vol. 26, no. 3, pp. 459–474, 1990.

[GAO 03] GAO Y., ER M.J., "Online adaptive Fuzzy neural identification and control of a class of MIMO nonlinear systems", *IEEE Transactions on Fuzzy Systems*, vol. 11, no. 4, pp. 462–477, 2003.

[GAS 01] GASSO K., MOUROT G., RAGOT J., "Structure identification in multiple model representation: elimination and merging of local models", *IEEE Conference on Decision and Control*, vol. 3, pp. 2992–2997, 2001

[GEN 72] GENTINA J.C., BORNE P., LAURENT F., "Stabilité des systèmes continus non linéaires de grande dimension", *RAIRO*, vol. 3, pp. 69–77, 1972.

[GRU 76] GENTINA J.C., GRUJIC L.T., BORNE P., "General aggregation of large scale systems by vector Lyapanov functions and vector norms", *International Journal of Control*, vol. 24, pp. 529–550, 1976.

[GUA 91] GUAN Y., SAIF M., "A novel approach to the design of unknown input observers", *IEEE Transactions on Automatic Control*, vol. 36, no. 5, pp. 632–635, 1991.

[GUE 04] GUERRA T.M., VERMEIREN L., "LMI-based relaxed nonquadratic stabilization conditions for nonlinear systems in the Takagi–Sugeno's form", *Automatica*, vol. 40, no. 5, pp. 823–829, 2004.

[HA 03] HA Q.P., NGUYEN H.T., TUAN H.D., "Dynamic output feedback sliding-mode control using pole placement and linear functional observers", *IEEE Transactions on Industrial Electronics*, vol. 50, no. 5, pp. 1030–1037, 2003.

[HAM 02] HAMMOURI H., KABORE P., OTHMAN S., BISTON J., "Failure diagnosis and nonlinear observer. Application to a hydraulic process", *Journal of the Franklin Institute*, vol. 339, pp. 455–478, 2002.

[HAS 99] HASSIBI A., HOW J., BOYD S., "A path-following method for solving BMI problems in control", *American Control Conference*, vol. 2, pp. 1385–1389, San Diego, USA, 1999.

[ISE 93] ISERMANN R., ULIERU M., "Integrated fault detection and diagnosis", *IEEE/SMC Conference on Systems Engineering in the Service of Humans*, vol. 1, pp. Le Touquet, France, 743–748, 1993.

[JAD 99] JADBABAIE A., "A reduction in conservatism in stability and L2 Gain analysis of T-S fuzzy systems via Linear matrix inequalities", *14th Triennial World Congress of IFAC*, pp. 285–289, Beijing, China, 1999.

[JIA 02] JIANGA B., WANG J.L., "Actuator fault diagnosis for a class of bilinear systems with uncertainty", *Journal of the Franklin Institute*, vol. 339, pp. 361–374, 2002.

[JOH 98] JOHANSEN M., RANTZER A., ARZEN K., "Piecewise quadratic stability for affine Sugeno systems", *IEEE International Conference on Fuzzy Systems Proceedings, IEEE World Congress on Computational Intelligence*, vol. 1, pp. 55–60, 1998.

[JOH 00] JOHANSEN T.A., SHORTE R.N., MURRAY-SMITH R., "On the interpretation and identification of dynamic Takagi-Sugeno Fuzzy models", *IEEE Transactions on Fuzzy Systems*, vol. 8, no. 3, pp. 297–312, 2000.

[JOH 03] JOHANSEN T.A., BABUSKA R., "Multiobjective identification of Takagi-Sugeno Fuzzy models", *IEEE Transactions on Fuzzy Systems*, vol. 11, no. 6, pp. 847–860, 2003.

[KAL 60] KALMAN R.E., BETRAN J.E., "Control system analysis and design via the 'second method' of Lyapunov-I: continuous-time system", *ASME Journal of Basic Engineering*, vol. 82, pp. 371–393, 1960.

[KOE 02] KOENIG D., MAMMAR S., "Design of proportional-integral observer for unknown input descriptor systems", *IEEE, Transactions on Automatic Control*, vol. 47, no. 12, pp. 2057–2062, 2002.

[KRU 08] KRUSZEWSKI A., WANG R., GUERRA T.M., "Nonquadratic stabilization conditions for a class of uncertain nonlinear discrete time TS fuzzy models: a new approach", *IEEE Transactions on Automatic Control*, vol. 53, no. 2, pp. 606–611, 2008.

[KUD 80] KUDVA P., VISWANADHAM N., RAMKRISHNA A., "Observers for linear systems with unknown inputs", *IEEE Transactions on Automatic Control*, vol. 25, no. 1, pp. 113–115, 1980.

[LEW 86] LEWIS F.L., "A survey of linear singular systems", *Circuits Systems Signal Process*, vol. 5, pp. 3–36, 1986.

[LIU 03] LIU X.D., ZHANG Q., "New approach to Hinf controller designs based on observers for T-S fuzzy systems via LMI", *Automatica*, vol. 39, pp. 1571–1582, 2003.

[LUE 64] LUENBERGER D.G., "Observing the state of a linear system", *IEEE Transactions on Military Electronics*, ME-8, pp. 74–80, 1964.

[MA 98] MA X.J., SUN Z.Q., HE Y.Y., "Analysis and design of fuzzy controller and fuzzy observer", *IEEE Transactions of Fuzzy Systems*, vol. 9, no. 1, pp. 41–51, 1998.

[MAR 99] TEIXEIRA C.M., STANISLAW H., "Stabilising controller design for uncertain nonlinear systems using fuzzy models", *IEEE Transactions of Fuzzy Systems*, vol. 7, no. 2, pp. 133–142, 1999.

[MOR 00] MORERE Y., Mise en œuvre de lois de commandes pour les modèles flous de type Takagi-Sugeno, Doctorate thesis, University of Valenciennes et du Haut Cambrésis, Lille, 2000.

[MUR 97] MURRAY-SMITH R., *Multiple Model Approaches to Modelling and Control*, Taylor and Francis, Londres, 1997.

[ORE 83] O'REILLY J., "Observer for linear system", in BELLMAN R. (ed.), *Mathematics in Science and Engineering*, vol. 140, Academic Press, New York, 1983.

[PAT 97] PATTON R.J., CHEN J., "Observer-based fault detection and isolation: Robustness and applications", *Control Engineering Practice*, vol. 5, no. 5, pp. 671–682, 1997.

[RIC 01] RICHARD J.P., *Algèbre et analyse pour l'automatique*, Traité IC2, Hermès-Lavoisier, Paris, 2001.

[SAI 93] SAIF M., GUAN Y., "A new approach to robust fault detection and identification", *IEEE Transactions on Aerospace and Electronic Systems*, vol. 29, no. 3, pp. 685–695, 1993.

[SHE 03] LIN S.F., WANG A.P., "Unknown input observers for singular systems designed by eigenstructure assignment", *Journal of the Franklin Institute*, vol. 340(1), pp. 43–61, 2003.

[TAK 85] TAKAGI M., SUGENO M., "Fuzzy identification of systems and its application to modelling and control", *IEEE Transactions on Systems Man and Cybernetics*, vol. 15, no. 1, pp. 116–132, 1985.

[TAM 96] TAMBE S.S., KULKOARNI B.D., DESHPANDE P.B., "Element of artificial neural networks with selected application on chemical engineering and chemical and biological sciences", *Simulation and Advanced Controls*, Louisville, USA, 1996.

[TAN 98] TANAKA K., IKEDA T., HE Y.Y., "Fuzzy regulators and fuzzy observers: relaxed stability conditions and LMI-based design", *IEEE Transactions on Fuzzy Systems*, vol. 6, no. 1, pp. 250–256, 1998.

[TAN 01] TANAKA K., HORI T., WANG H.O., "A fuzzy Lyapunov approach to fuzzy control system design", *IEEE American Control Conference*, vol. 6, pp. 4790–4795, 2001.

[TAN 02] TAN C.P., EDWARDS C., "Sliding mode observers for detection and reconstruction of sensor faults", *Automatica*, vol. 38, pp. 1815–1821, 2002.

[TUA 01] TUAN H.D., APKARIAN P., NARIKIYO T., YAMAMOTO Y., "Parameterized linear matrix inequality techniques in fuzzy control system design", *IEEE Transactions of Fuzzy Systems*, vol. 9, no. 2, pp. 324–332, 2001.

[UTK 92] UTKIN V.I., *Sliding Modes in Control Optimization*, Springer-Verlag, Berlin, 1992.

[VAL 99] VALCHER M.E., "State observers for discrete-time linear systems with unknown inputs", *IEEE Transactions on Automatic Control*, vol. 44, no. 2, pp. 397–401, 1999.

[WAL 87] WALCOTT B.L., CORLESS M.J., ZAK S.H., "Comparative study of nonlinear state observation", *International Journal of Control*, vol. 45, pp. 2109–2132, 1987.

[XIO 03] XIONG Y., SAIF M., "Unknown disturbance inputs estimation based on a state functional observer design", *Automatica*, vol. 39, pp. 1389–1398, 2003.

[XU 93] XU L.., KROYZAK A., ORJA E., "Rival penalized competitive learning for cluster analysis RBF and curve detection", *IEEE Transactions on Neutral Network*, vol. 4, pp. 636–646, 1993.

[YAN 88] YANG F., WILDE R.W., "Observers for linear systems with unknown input", *IEEE Transactions on Automatic Control*, vol. 33, pp. 677, 1988.

[ZAD 65] ZADEH L., "Fuzzy sets", *Information Control*, vol. 8, pp. 338–353, 1965.

Index

P, R

S

T, U

www.ingramcontent.com/pod-product-compliance
Ingram Content Group UK Ltd.
Pitfield, Milton Keynes, MK11 3LW, UK
UKHW020818290926
15470UKWH00003B/1226

9 781848 214125